Women's Rights, Racial Integration, and Education from 1850–1920

Women's Rights, Racial Integration, and Education from 1850–1920

The Case of Sarah Raymond, the First Female Superintendent

Monica Cousins Noraian

WOMEN'S RIGHTS, RACIAL INTEGRATION, AND EDUCATION FROM 1850–1920
Copyright © Monica Cousins Noraian, 2009.

All rights reserved.

First published in 2009 by
PALGRAVE MACMILLAN®
in the United States—a division of St. Martin's Press LLC,
175 Fifth Avenue, New York, NY 10010.

Where this book is distributed in the UK, Europe and the rest of the world, this is by Palgrave Macmillan, a division of Macmillan Publishers Limited, registered in England, company number 785998, of Houndmills, Basingstoke, Hampshire RG21 6XS.

Palgrave Macmillan is the global academic imprint of the above companies and has companies and representatives throughout the world.

Palgrave® and Macmillan® are registered trademarks in the United States, the United Kingdom, Europe and other countries.

ISBN: 978–0–230–61322–5

Library of Congress Cataloging-in-Publication Data

Noraian, Monica Cousins.
   Women's rights, racial integration, and education from 1850–1920 : the case of Sarah Raymond, the first female superintendent / Monica Cousins Noraian.
      p. cm.
   Includes bibliographical references and index.
   ISBN 0–230–61322–5
    1. Raymond, Sarah, d. 1918. 2. School superintendents—United States—Biography. 3. Women school superintendents—United States—Biography. 4. Women's rights—United States—History. I. Title.

LA2317.R316N67 2009
371.2′011092—dc22   2009013888
[B]

A catalogue record of the book is available from the British Library.

Design by Newgen Imaging Systems (P) Ltd., Chennai, India.

First edition: November 2009

10 9 8 7 6 5 4 3 2 1

Printed in the United States of America.

CONTENTS

Acknowledgments		vii
One	Introduction	1
Two	The Early Years	19
Three	The Illinois State Normal University Years	35
Four	Teacher and Principal of Bloomington Schools	67
Five	Superintendent of Bloomington Schools	83
Six	The Resignation	113
Seven	Leading beyond the Schools: Community Involvement in Bloomington, Boston, and Chicago	125
Appendix	Literature Review	135
Notes		153
Bibliography		171
Index		185

ACKNOWLEDGMENTS

> For most of history, anonymous was a woman.
> —Virginia Woolf

I dedicate this to those who, like Sarah Raymond Fitzwilliam, were inspired by others to challenge themselves and society. She believed in herself, the goodness of others, and the potential for a better community. May we all find that person who inspires us to make our world a better place and live our lives with compassion for all. History needs to be *his story*, *her story*, and our story. Here is to giving voice to an otherwise anonymous woman in history.

Thanks to the support, interest, and hard work of others, Sarah Raymond's voice and legacy is now being heard again. Her story is inspiring and I hope it will continue to empower others who read it. We can make a difference in our community and challenge injustice. I was lucky to find a woman whose story needed to be told and I am even luckier to have others who empowered me with my research on Sarah Raymond to challenge myself, our program on educational leadership, the early history of Illinois State University, and the community and schools of Bloomington/Normal, Illinois.

I appreciate them for listening to my concerns, engaging my questions, fostering my voice, challenging my paradigms, and celebrating my passion and spirit. They are an amazing group of female scholars and I am proud to call them my mentors. Thank you Dr. Amee Adkins, Dr. Sandra Harmon, Dr. Linda Lyman, and Dr. Trisha Klass.

I thank my husband, my children, my parents, my extended family, my friends, and my history department colleagues for believing in me, for challenging me, for encouraging me, for helping me, for loving me, and for modeling for me a life of service and dedication to social justice. Thank you for empowering me to challenge myself and those around me.

To my circle of friends who keep me grounded, laughing, and in touch with reality—Amy, Vicki, Liz, Heather, Susan, Molly, Tammy, Nancy, and Emily, you are the sisters I never had. To my circle of friends at work—you make life, teaching, and learning a pleasure. Thank you for supporting me, challenging me, and encouraging me to image doing things differently.

A special thanks to my mom, Elaine, for the proofreading, editing, indexing, and countless hours of support over the years. I know I am where I am today because I had a mom who believed in me and was there for me. I only hope my girls grow up knowing the same thing and that I can be there to support their dreams no matter how big.

To my dad, Peter, who was the first to suggest I get a PhD in History/Social Foundations. I continue to be inspired by his spirit of historical inquiry, accuracy, and justice. He dedicated his life to interpreting history and bringing it alive for others. He made the ordinary objects of the past become extraordinary windows into our history.

I am very appreciative of the various archives and archivists who welcomed me and facilitated my research. The work of historians cannot be completed without the support of others who know their sources and preserve the memories of the past. At the McLean County Museum of History, I thank Greg Koos and Bill Kemp. At the Illinois State University Archives, I thank Dr. Jo Rayfield and I also thank the librarian staff at the Illinois State University, Vanette Schwartz and Sharon Naylor and Illinois Wesleyan University. Thank you to the support at the Kendall County Historical Society and the Chicago History Museum/Archives. A special thank you and dedication to Bloomington School District 87 who had the vision to keep historic records/documents and are eager and honored to celebrate the rich past of their school district. And many thanks to the friends and colleagues like Dr. John Freed who thoughtfully reviewed the manuscript and shared their perspectives.

I appreciate my graduate student Sara Piotrowski who read books beyond her interests, traveled to research sites beyond her imagination, and offered support and friendship beyond her knowing. The editorial staff at Palgrave Macmillan has been fabulous and made the challenging tasks more manageable; so thank you to Samantha Hasey, Julia Cohen, Allison McElgunn, Maran, and Vidhya.

This is dedicated to those who influence others without even knowing it.

And last there are few works that truly capture the love and appreciation I have for you my husband Kirk and my daughters Emma and Celeste. You spent many weekends without me as I finished my writing; I truly appreciate all that you did to help this dream become a reality. I can now answer the nightly question of "Are you done with your book yet?" (That rang like the familiar call of Are We There Yet?). Yes, I am finally done with my book.

M.C.N.

CHAPTER ONE
─────────────
Introduction

> Biography is history made personal.
> —Alan Wieder, *Writing Educational Biography*

How do we know what we know about history? Who tells the stories of the past? Whose stories do we hear and whose voices are underrepresented or even lost? Whose contributions get noticed and whose go unrecorded or forgotten? Millions of lives pass without attention or record. Yet their experiences are real and their impact on others profound. What is the difference between those people with notice and those without? In most cases it is that someone cared enough to trace their story, record their findings, and share them with others. When is a life untold worth telling? It is a lucky historian who finds a story worth telling and a lucky reader who becomes engaged in the lost voices of the past.

Sarah Raymond is one whose story has, for the most part, been untold. Who was Sarah Raymond, or Sarah E. Raymond Fitzwilliam to be specific? From the records available, I will develop her portrait, record her voice, and renew the power of her legacy so we may learn from this remarkable woman and the contributions she made. How did one woman learn to dare and then dare to lead others?

Essays by women biographers help to shape the questions and the consciousness of this book and to liberate the historian. They talk of a responsibility to justly and generously honor their subjects.[1] At the same time the authors talk about recognizing a power and silent mending that transpired for both the author and the reader about a forgotten subject. This new field calls for new approaches to the writing and interpretation of women's voices, which time continues to obscure and silence.[2] There is a present and future being transformed by the writing of feminist biography and an ability to show the interaction between women and social structures that are dominated by men. "By creating a history of women, historians do more than reconstruct the past in new ways. They transform the possibilities in women's present and future."[3]

Kathleen Weiler reflects on the role of historian and challenges the traditional notion of organizing and presenting a story about the past and suggests a more feminist perspective, including and making more overt the motives and theoretical and political assumptions that guide historical narratives. She cites examples from Shulamit Reinharz and Carolyn Steedman who connect personally to the research through intellectual questions and personal reflections by understanding and articulating the researchers concerns that shape the presentation of the past. Kathleen Weiler introduces her historical narrative by sharing her own personal and intellectual concerns and nicely bridges the recent scholarship of women's history and feminism.[4]

When I began this study, I knew only that there was a Sarah Raymond School, that Sarah Raymond had been an early superintendent of Bloomington, Illinois, schools, and that she had published an important book on curriculum and instruction. I was drawn to her right away as I, too, am a teacher and historian, and felt my interest in educational leadership and the work of women in schooling connected well with her story. As I uncovered more, I realized how groundbreaking her story was and how important restoring her place in history would become. The pursuit of the story was as exciting as the story itself, which makes for a great read and a great write. Challenging the injustice of race and gender, establishing professionalism in schools, and breaking the boundaries for women in leadership are a few of the fascinating stories that come out of this biography of one woman's voice and choices.

She is buried in Bloomington's Evergreen Cemetery and her tombstone leaves a permanent record of her name. Also standing as monuments to Sarah Raymond are the Sarah Raymond School and a cornerstone in Withers Park with her name carved in as library board president. She was the subject and author of articles in local newspapers. Pertinent documents reside in the McLean County Museum of History and the Bloomington District 87 school district archives. The quest was on. I became convinced that her story contributes to larger questions of historiography, women in educational leadership, and social justice. Uncovering the story of Sarah Raymond and writing her biography offer a significant contribution to the lost voices of women in educational leadership and the early history of Bloomington-Normal and the Illinois State Normal University (ISNU).

This story begins at Evergreen Memorial Cemetery. The cemetery was originally two separate ones adjacent to each other. Bloomington Cemetery was founded in 1850 funded by city tax dollars, while the other called Evergreen Cemetery was established in 1860 and privately funded.[5] In the 1960s they were joined and have remained that way. There are several notable persons buried in Evergreen Cemetery: David Davis, an associate justice of the U.S. Supreme Court; Charles Radbourn, Major League Baseball Hall of Famer; Adlai Stevenson I, Vice President of the United States; Adlai Stevenson II, U.S. Ambassador to the United Nations, Governor of Illinois, and presidential candidate; and Carl Vrooman, assistant Secretary of Agriculture under Woodrow Wilson. In an attempt to curb cemetery vandalism and to celebrate local history, the McLean County Museum of History began organizing cemetery walks there each year, highlighting stories of the famous and ordinary people interned there. Sarah Raymond was a "voice from the past" cemetery walk character in 1996.[6]

Her grave marker is flat to the ground and can be found in section 9 of the cemetery. It simply reads: "Sarah E. Raymond

Fitzwilliam Died Jan. 31, 1918." She is buried in the Trotter Family Plot. Georgina Trotter was a long-time friend and colleague of Sarah Raymond who died in 1904. She too was highlighted in the *Voices from the Past* cemetery walk.[7] Across the path in section 10 is the Fitzwilliam Family plot. Intriguingly, Sarah Raymond is buried neither with her husband, F.J. Fitzwilliam, who died in 1899, nor with her parents. Her father, Jonathon Raymond, died in 1884, and her mother, Catherine, in 1877. Her parents are buried in section G-3 of the Evergreen Cemetery, just two sections north of where Sarah Raymond lies.[8] Her burial with her very close companion Georgina Trotter leads to further questions and areas of research.

Who Was She?

The following chapters will unfold Sarah Raymond's life as the biography explores her childhood, schooling, and early teaching in central Illinois, her years at the Illinois State Normal University and her work with the Bloomington, Illinois, school district as teacher, principal, and city school superintendent. It will explore her continued relationship with the ISNU and her later involvement socially, civically, and culturally through her work with various women's clubs and organizations. The biography investigates factors that allowed her to break boundaries and roles for women and poses the question how she learned to dare and dare to lead. Overall it will show how she was ahead of her times as a precursor to the "new woman" movement by illustrating her professional work, her relationships, and her social involvement. She was an active professional woman, living in Bloomington, Boston, and Chicago. She married later in life and bore no children of her own.

How Is She Remembered?

Obituaries are a capsule of information about a person's life. I found six from various sources across the county. That she was included in the *Voices from the Past* and received significant coverage in the press of her day indicate that she was a person worth studying and with a voice worth reporting. My focus turned to the woman behind the obituary. Piecing a life's story together from obituaries leads to a series of challenges: who wrote the piece, how well did they know the woman, what was their connection to her? History is recorded but is only as accurate as its writers and the sources they use. I encountered inconsistencies about her life story in the various sources, but there is enough significant evidence to support her place in history. As a historian, I began to tackle the stack of obituaries about Sarah E. Raymond hoping to uncover the woman, her life, and her choices. I was ready for the next layer of story and what I would learn from these obituaries. Decoding and deconstructing these various accounts of her life proved to be challenging. Such an analysis is the important work of historians: following up on leads, confirming truth and error, and telling a story left un-narrated. What comes out of these obituaries is interesting. There are the facts, omissions, and errors of her professional career, social activism, and personal information.

There is what is reported and what is left out, intentionally or unintentionally. Thorough or short, it is clear that there is a "politics of voice" with regard to the story being related. Are papers willing to take a stand and give recognition to an outstanding woman ahead of her time? Are they comfortable promoting the contributions of women in a male-dominated society? Do they focus their report on safer, more comfortable topics rather than highlighting more socially challenging information? The life of Sarah Raymond is rich in professional, social, and personal achievements; but the obituaries vary on their level and selection of information.

The shortest obituary was published in *The Chicago Evening Post* and reported Mrs. Sarah Raymond Fitzwilliam was "first woman superintendent of schools in the country."[9] She was significant not only as a Chicago resident, where she lived at the time of her death, but also as a woman ahead of her time. This obituary in a prominent national paper quickly established her noteworthiness and points to her professional prominence. This is a strong statement of fact.

The *Chicago Daily Tribune* reported, however: "She is said to have been the first woman superintendent of schools in the country."[10] She was noted in the Chicago paper for her local residence, and more importantly she was selected for inclusion in the paper for her reported noteworthy professional status. There is an interesting tone difference between the two Chicago papers. The language of the *Post* was stronger, reporting she was the first woman superintendent of schools in the country. The *Tribune*, on the other hand, left that question open. The *Tribune* obituary is also brief, but significant since it, a major newspaper, carried the story and took a stand on her professionally as a prominent woman in educational leadership and school administration. This paper went a bit more in depth about the contributions of Sarah Raymond. The obituary noted that she had served as regent of the Daughters of the American Revolution and held the role of historian and that she had been an enthusiastic art collector.

However, three pieces of information from this short paragraph obituary proved to be incorrect. *The Chicago Daily Tribune* inaccurately reported that she was survived by her husband Frank Fitzwilliam. Her husband, who passed away in 1899, preceded her in death by over ten years. He is referred to as Frank in this report but other references identify him as Captain F.J. or Francis Julius. Although correctly noting that Sarah Raymond had been the superintendent, the *Tribune* misrepresented the years of service by saying "serving as head of the Bloomington, Ill. schools from 1865 to 1885." She was actually connected with the schools as teacher and administrator from 1868 to 1892, and served as superintendent from 1874 to 1892. The third inaccurate report was regarding her involvement with the Art Institute of Chicago (AIC). The *Tribune* reports that Sarah Raymond "willed several noted paintings to the Art Institute." Communication with the AIC Archives reveals interesting information. Mrs. Fitzwilliam was a Life Member of the AIC and paid one hundred dollars to join. The AIC has in its permanent collection seventy-seven pieces given by Mrs. Fitzwilliam. None of the pieces are currently on display and only ten of the pieces are photographed. All of the works were donated in 1917, a year before her death in 1918. The items are currently housed in two departments: Asian Arts and Textiles (items donated include Hindu drawings and textiles) and Decorative Arts (items donated include cups, plates, paperweights, bracelets, cameos, etc.). According to Marie Kroeger of the AIC Archives, the AIC shows no record of Mrs. Fitzwilliam willing several noted paintings, but it is clear that she was an enthusiastic art collector and supportive member of the Institute.[11]

Being included in the paper of a major city indicates importance. Each paper set the obituary in a position of distinction. Her death was also noted in The *Journal of the Illinois State Historical Society* and included both personal and professional recognition of Sarah Raymond. Interestingly this is the only obituary with an included author, Charles Capen. Does that give it more credence? Charles Capen was associated with the ISNU as a member of the University High School's first graduating class in 1865. He was a Harvard graduate who practiced law in Bloomington. Later he was a member of the Normal School Board. He writes about Sarah Raymond with a sense of connection, personal knowledge, and pride. There is a clear level of respect and admiration for her work and contributions to society. He described her as follows: "Untiring, self denying, able and tactful, she gained a national reputation, the memory of service, and the enduring gratitude of her contemporaries and of future generations."[12] He continued "More than of most others it may be said her good and useful works do follow her; she spent her life for others, and the State is better for her having lived in it. She well deserves permanent recognition in our annals."

I could not agree more with the eloquent words; however, I am not sure how strong her legacy is today. It saddens me to hear how optimistic and self-evident it seemed at her death that Sarah Raymond would remain nationally recognized and how forgotten her story and achievements have become. At the same time, it is empowering to know that this woman should be in our texts and our historical memory and will again regain her rightful place with others who dared, challenged, and stood ahead of their time.

When we uncover the many organizations with which she was connected, it is evident that she worked untiringly. When one comes to understand the levels of leadership she attained, it is clear that she was able, and when one thinks of the politics she had to negotiate to rise to prominent positions in both social and professional organizations, her tact seem evident. She was appreciated by her contemporaries and will be equally embraced by modern readers of her story when they come to discover her fully.

In the era where she came into power and influence, it was not typical for women to devote themselves beyond their husbands, families, and communities. Therefore professional woman did not marry, and married women with children were not able to take on the public roles of their single colleagues. Women had a choice to marry or remain "professional."

Sarah Raymond was a servant of the community as teacher, educational leader, and active club woman. She was self-denying to her own needs as indicated in the obituary and exemplified in a letter she wrote to her brother.[13] She begins the letter, "I thought I would write you a few lines to night though I am very tired and do not think I shall say anything very brilliant or entertaining. Should write often if I could possibly find the time but you can hardly realize how much I have to do." She continues with the idea of buying land for financial self-sufficiency.

> I think I should like to buy a piece of land near some growing village and let it rise in value as a means by which to make my fortune for it is very evident that I have my own to make. No smiling lad comes knocking at my door, bringing with him the deformities of a well stuffed purse to deliver at my feet to will and do with as my good pleasure.

She ends the letter with:

> I have no new dresses cloaks hats nor bonnets so I have nothing of woman's gossip to write you. I am living on an economical

scale. Wearing out my old clothes which due to my previous economy are not an overstock. The ladies here are dressing very much this winter but being a Yankee school marm do not come in the 1st circle.

Signed "I remain your affectionate sister, Sarah."

Another letter in the file, also undated and cowritten with her father Jonathan, reiterates how busy she is:

wills are all exhausted making preparations for examination which comes off this week. I have been very busy this term keeping up with my class and making up the lessons which I lost before entering school besides this I have to walk about 2 ½ miles you know our folks are always fearful lest I shall not have exercise enough. Dame Nature has assumed a terrific face today and I do not like to make my explorations extend a very great distance...[14]

She is described in terms of her day, revealing perhaps more about the values of a great leader in the late 1800 to early 1900s. It is clear from the descriptions that her contemporaries valued and respected her and felt her legacy would be long revered. I look forward to securing her place in the annals of educational leadership.

The Alumni Quarterly of the I.S.N.U. includes another obituary for Sarah Raymond. The ISNU was among the premier teacher training institutes in its day. The *Quarterly* reported that "In her death the Alumni Association loses one of its most famous members."[15] There is a unique introduction and then a reproduction of the Bloomington *Pantagraph* obituary dated February 1, 1918. A strong sense of value and respect for Sarah Raymond and her accomplishments was shared in the article. "A list of the positions of responsibility she has held is mentioned in this number of the quarterly, but no publication will ever be able to state the enrichment which came into the lives of thousands because of her work." Similarly interesting thoughts and values are shared about ISNU: "Her life affords a splendid example of the inability of the statistician to compute the good that is done to society by such an institution as the school which claims her as a member of its alumni." Continuing in that vein

> An expert accountant can consult the records of the ISNU and compute within a few dollars the cost to the state for the four years schooling which this institution gave to Miss Raymond, but the accountant does not exist who can tell the worth to society of the results of the inspiration which she received as a student here.

There is a strong sense of the interconnections between the state, the university, and teachers and the priceless and powerful impact of education. "The state, through the University, invested a few talents in her education, and she returned that investment a thousand fold."[16]

The Pantagraph ran the obituary heading and following subheadings: "Mrs. Fitzwilliam dead in Chicago, former superintendent of Bloomington schools, end of a notable career in educational work in this county—originated Trotter Fountain." A photo is included as well with the caption "Useful Life Ended."[17] *The Pantagraph* also made some generalizations and errors in its reporting of the death of Sarah Raymond. They added a hyphen to her name "Mrs. Raymond-Fitzwilliam," cited her years of service as superintendent as "several," and misspelled the "Illinois Platro Club," which should read Plato Club, of which she was president for two years.

The Daily Bulletin of Bloomington seemed to downplay the number of years Sarah Raymond was connected with the Bloomington schools and superintendent by reporting: "For a number of years Mrs. Fitzwilliam was superintendent of the city schools of Bloomington."[18] It is certainly not inaccurate, but it seems to minimize the eighteen-years she spent as superintendent.

It is clear that *The Daily Bulletin* and *The Pantagraph* with their lengthy articles held Sarah Raymond in high regard for her contributions to the community and the schools. The six obituaries clearly support the position that she was a well-respected educational leader whose place has been lost in our annals today. *The Pantagraph* and *The Daily Bulletin* obituaries also give extensive discussion of her early life. Her personal and professional undertakings will be discussed in later chapters.

Little is known about early women in educational leadership. Most early coverage is anecdotal, superficial, and inconsistent. The recent literature in the field is much stronger but focuses on a later period than the years of Sarah Raymond and her superintendency. This both challenges and excites the researcher as this is a much needed case study. Women's place in early education has for the longest time been recorded in folklore and societal memory more than in the annals of scholarly research. Although the stories are real to the families telling them, they are just that—stories of the past. I wish to claim these voices for scholarship and to record the experiences and lessons for future generations.

The biography of Sarah Raymond is larger than the reclaiming of one woman's voice. It is the writing into our history the significant contributions and monumental first steps toward equality and opportunity for all. The work of the nation's first woman superintendent should be widely known and understood. It should be easy to find and widely included in reference texts on the history of education. There are universal lessons to be gained from understanding this first step in the rise of women to positions of leadership, and scholars have been remiss by excluding them. With the voices recorded and stories understood there should be no reason not to.

Sarah Raymond Fitzwilliam is a familiar name in the Bloomington-Normal community because there is a school with her name. The unique qualities of her life and school leadership are not familiar, however.[19] Her legacy is folklore in the local community, and does not extend far beyond that. In fact, the local newspaper in its daily local trivia section, recently posed the question "Can you name the Central Illinois city that is believed to have had the first female school superintendent in America? (a) Streator; (b) Stanford; (c) Urbana; (d) Clinton; (e) Bloomington." The paper shares as the answer "It was in (e) Bloomington, where in 1874, Sarah E. Raymond (and later namesake of Raymond School) was appointed superintendent of schools after serving as principal of Bloomington High and also principal of the No. 5 School (later Bent). She was superintendent until 1892."[20] How many other isolated stories are there of powerful women left out of scholarly texts across the country? There needs to be an effort made to write these women into the history of educational leadership, just as was done for other voices underrepresented in history. The challenge is making that happen.

Today it is easier than ever to discover local popular memory through the Internet. Google-ing, to use popular vernacular, is a new way of historical discovery, which must be followed up with more traditional research. It opens new access to history and localized stories. The liberation of knowledge and information has led to a new politics of knowledge and power. Sources left out of primary texts no longer need to remain voices lost. They now have an outlet in the new grassroots movements of local popular memory. Although credibility and accuracy of sources remains an important agenda for historians, the type of work they do has changed. The new politics of knowledge embraces truthfulness rather than the truth. This new way of thinking and discovery about truth is critical to the rise of new ideas and areas of historical thinking.

Barbara Finkelstein and Craig Kridel in their work with biography and educational writing explore the fact that biography as a form of historical research has a long tradition even though such inquiry within the field of educational history is not extensive. They suggest biographical research, in one sense, becomes an alternative to the grand historical interpretations of education. Barbara Finkelstein writes: "Through the lens of biography historians have constructed creative windows through which one can glimpse several otherwise undiscoverable realities. Indeed, biography constitutes a unique form of historical study that enables education scholars to explore intersections between human agency and social structure."[21]

Biographical studies offer, according to Finkelstein and Kridel, four entrees into the study of history. First, they suggest, biography is a lens through which to explore the origin of new ideas. Second, biography offers a window on social possibility. Third, biography provides a way to view relationships between educational processes and social change. And fourth, it is a form of "mythic overhaul"—a way to glimpse the variability and complexity of life within a single era or over longer periods.

The biography of Sarah Raymond sheds light on her life and times and offers insight, examples, and perspective on larger issues of women in nineteenth-century American society.

The following are points of significance with regard to the life and work of Sarah Raymond Fitzwilliam: (1) It is an important Central Illinois history story: graduate of ISNU, teacher, community activist, and an individual who made a difference in the community and the lives of others. (2) She was known by prominent people of the day. (3) She is a fine example of a historic woman in leadership. (4) She is a pre-progressive educator and reformer. (5) She was a leader in curriculum development and an example of teacher leadership/professionalism. (6) She challenged gender and race discrimination in the schools and advocated equal pay for equal work and integration. (7) She held leadership roles in educational professional organizations: president of the Woman's Educational Association of Illinois Wesleyan University, twice president of the Woman's State Teachers Association, acting president of the Central Illinois State Teachers Association, secretary of the Illinois State Teacher's Association, and first president of the School Mistresses' Club of Illinois.

The personal and public journey of Sarah Raymond illustrates nicely the course of women's leadership in American society during the Gilded and Progressive eras. The organizations she supported and the causes she advocated follow a path of reform toward social justice. Women's rights, women's suffrage, and American reform are related, intertwined, and dynamic movements. The path toward change is steadfast but the road is varied following society's changing needs. There are three stages in the evolution of the woman's suffrage movement. The pre–Civil War movement was focused on universal reform, during the Reconstruction era the movement was scattered and independently organized, and after 1885 the suffrage movement grew into an organized national movement.[22] It is interesting that during this time the suffrage movement followed parallel challenges of American reform. At times social activist were uplifted by their faith, at other times confronted by narrowed potential to change fundamentally the nature of American society. Women gained rights and ultimately changed the nature of American society by uniting with other groups and linking up with other reform causes.

In the 1820s universal education was seen as a way for women to gain access and the ability to participate in democracy. Moral reform was another social movement that united women. Women participated in the antislavery struggle. By the 1840s, however, the role of women in the movement was controversial. Women were gaining experience, self-confidence, and a voice with the movement. They were also feeling oppression empathy. They learned how to organize and speak up about oppression and injustice, and it was these skills that forged a larger women's rights movement.

The next generation of woman's rights supporters came in the 1850s as the political system opened up and the individual triumphed. Natural rights, moral responsibility, and individualism were the new tenets of politics and women were ready to seize the opportunity for claiming their rights. Women like Lucretia Mott and Elizabeth Cady Stanton had a cosmopolitan sensibility, a broad education, and a fundamental mistrust of human institutions.[23] Women were redefining themselves and their dress. The Bloomer was the new American costume and was causing a stir. Conventions, public speaking, and the press were their methods. They had neither political power nor national controversy until they united with temperance agitation.

By the 1860s women were seeing small victories and their own movement was gaining momentum. Educational opportunities were expanding and women were entering other professions. They were gaining property rights and discussion of woman's rights was up for public scrutiny. The challenge was to make this a national movement with political power. The Civil War would do just that. Women were now important allies to the Radical Republicans. They were gaining political esteem and power. New organizations formed (The Woman's National Loyal League) to give the movement what it lacked before the war. Women were uniting in support of larger issues. An American Equal Rights Association was founded in 1866. Then in 1869 the National Woman Suffrage Association was created and then shortly in response to that the American Woman Suffrage Association was formed. In 1890 the National American Woman Suffrage Association was founded. The problem was in uniting the country toward reform. The issues and the organizations were not strong enough and the American nation too resistant to change. Reformers again held to their goals and again there were many different paths. Small victories came. Wyoming granted women the vote in 1869, Utah in 1870, and school suffrage extended on a state-by-state basis. Illinois women obtained the vote in school board elections in 1892. The suffrage press continued to publish but the main dynamic in women's reform activities of the Gilded Age lay outside the woman suffrage movement. Most women felt their service was to their community and family. The Civil War had helped to define woman's capacity to organize and contribute, but more women participated in the temperance, social purity, and the suppression of vice movements than in suffrage.

The scale of women's involvement grew during the 1880s and 1890s. They were active with charities, churches, temperance, self-education societies, and the suffrage movements and other organizations grew. These "new women" professional, independent-minded, middle-class women were expanding the reform movements. More than in earlier years, these moral reformers saw men as the blame for society's ills. At the end of the 1880s, the Woman's Christian Temperance Union (WCTU), women's clubs, and American Woman Suffrage Association (AWSA) leaders joined in endorsing the ethical socialism of Edward Bellamy's 1888 novel *Looking Backward*. In the novel women are in a stronger position of equality to men; women receive equal pay for equal work and are wardens of the world.

Just as Sarah Raymond was doing, these "new women" were not just talking about issues in clubs but were acting on them as well. Sarah Raymond challenged issues of opportunity, equality, and pay for women. These "new women" opened homes for the poor and disenfranchised and prepared for greater political opportunity. Interestingly, historians Mari Jo and Paul Buhle identify two areas of concern regarding women's suffrage. Catholics feared that the women's vote would be used as a weapon of nativism and that white women's votes would guarantee control over the black population.[24] These very controversies played out in Bloomington in 1892 when women voted for the first time in school board elections. We will see a rise of nativism, concern over Catholic versus Protestant, and black women protesting the vote.

By the turn of the century numbers grew as did the alliances. Progressivism pushed reform through political means. Young college women joined the movements, middle-class urbanites, socialists, and laborers joined, strengthening the appeal to the working class. Unity and mass publicity would help change the future of suffrage. Women gained the electoral vote in 1913 in Illinois through a coalition of the National American Woman Suffrage Association (NAWSA) lobbyists and Progressive party officials in the state legislature. A thousand Illinois saloons were closed by municipal elections. After that other states prevented women from voting by "invisible forces" of liquor dealers and machine politicians—but also, no doubt, by thousands of voters who elevated their right to drink over woman's right to vote.[25] Formed in 1890, NAWSA was the result of a merger between two rival factions—the National Woman Suffrage Association (NWSA) led by Elizabeth Cady Stanton and Susan B. Anthony, and AWSA led by Lucy Stone, Henry Blackwell, and Julia Ward Howe. America's entry into World War I again solidified the suffrage movement. Women supported the country through peace efforts and with Woodrow Wilson's belated support, woman suffrage at last passed through Congress as a war measure. Sarah Raymond like so many other women in her day challenged society—by word and deed. She lived her life exemplifying the struggle for women's rights and died as that major step, the passing of the nineteenth amendment, was finally being realized.

CHAPTER TWO
===

The Early Years

> Kendall County has good reason to be proud of one of its early residents, Mrs. Sarah E. (Raymond) Fitzwilliam, whose life history is an example to the younger generation in this and other sections.
> —J.H. Burnham, Illinois State Historical Society

Where does one learn about the early years of someone's life? Why and how is it significant to their later accomplishments? How do we know what we know about why people are who they are? It is an interesting challenge to assemble the pieces of a puzzle and an even more challenging task to create a portrait of a person capturing the whole of their life story. One might not be able to do it all but one can develop the contextual picture.

Early life experiences help shape the adult. In the case of Sarah Elizabeth Raymond Fitzwilliam, I have found her own writings, both published and unpublished, as well as small biographies about her from various sources and time periods. These varied perspectives on early experiences help inform the reader and shed light on her choices later in life. Sarah Raymond was significant as an educational leader later in life, and it was those early experiences and the larger context of her formative years that allowed her to challenge segregation of students as a building principal, gender roles as superintendent, and societal norms as a community activist.

Sarah Raymond Fitzwilliam looked for social justice as a leading woman ahead of her time in great part perhaps because of the lessons she learned from her parents and earlier generations. Reportedly, her parents, Catherine and Jonathan Raymond, were early abolitionists who were said to have led over two hundred slaves to freedom through their house, a stop on the Underground Railroad. Her father, Jonathan Raymond, was a descendant of one of the early settlers of Ipswich, Massachusetts. William Raymond, who joined New England's famous rebellion against Governor Edmund Andros in 1687, was imprisoned at Boston, and thus became one of the early defenders of American independence. The town seal of Ipswich bears this inscription: "The birth place of American Independence, 1687."[1] J.H. Burnham, the first director of the Illinois Historical Society and the author of the biographical sketch of Sarah Raymond Fitzwilliam in the *History of Kendall County*, mentions an interest in her personally because his ancestor was also imprisoned with William Raymond in Boston and the Raymonds were family friends of his wife's family from Kendall County. Sarah Raymond attributes much of her parents' character to her grandparents and perhaps her own character to them as well.[2] It is interesting to see that many of the words chosen to describe her family fit her also. The author of the biography found in the *McLean County History* described her mother's and father's ancestors thus:

> Her mother descended from a stock which traces back to persons of historic distinction and whose representatives are noted for their success as thinkers, lawyers and business men. Her father found the fountain head of his ancestors in noble commanders and organizers. His immediate predecessors were especially distinguished in educational and religious work.[3]

Sarah Raymond described her grandfather William Raymond as a "man of ability, highly respected and for his superior goodness of character and was known in the community as 'Deacon.'" Her grandfather was a farmer and cooper but died when her father was about fourteen years of age. Her father apprenticed to learn mechanics and he developed "a remarkable talent and genius in this line" and helped to build the first cotton mill in Lowell, Massachusetts. Her paternal grandmother's family was "among the leading public men of their day and especially noted for their thorough Christian character." Her grandmother was a "fine conversationalist and was noted for doing good to all who needed her attentions."[4] There seems to be a value placed on good character and helping others.

Sarah Raymond describes her father Jonathan as being "endowed by nature with a powerful physical frame, coupled with the great strength of will, decision of character, and tempered with deep convictions of the beauties of justice." He was also a "good story teller of close observation, retentive memory, great force of character, strong will—was in some cases even austere, in others deeply sympathetic—of commanding figure and powerful voice."[5] *The Kendall County History* biography notes "his principles were deeply laid and vigorously maintained. His reputation was spotless, his integrity unblemished, his life pure and upright, without ostentation." Sarah Raymond grew up in a family of valorous tales, strong values, beliefs, and convictions. It is not surprising then that she, too, would be a model in the community. As sheriff from 1856 through 1858, her father was successful at catching horse thieves and broke up an organization known as the "bandits of the prairie." According to Art Thompson, whose grandfather was Sarah's brother Lyman and owner of the family papers, letters, and unpublished histories, Jonathan Raymond's picture remains on the wall in the Kendall County Sheriff's office.[6] He was also helpful to his neighbors. In 1852, with the local epidemic of cholera, he reached out to help those whom the physicians had abandoned due to the severity of the epidemic.

Sarah Raymond wrote a biographical piece about her mother for *The History of Kendall County*. Catherine Holt Raymond was among the "pioneer women who found a place in the high regard of the community in which they lived." Sarah Raymond also noted that "any account of social or political movements which ignore the part played by the women is incomplete, and Mrs. Raymond belonged to the class of her sex which always did much to cooperate with that of men." Sarah Raymond also writes:

> A striking element of her character was her readiness for any emergency and she was equal to all occasions. A woman of rare judgment, she possessed strong legal and scientific tendencies, but living retired from the world, the incidents of her life were domestic and ordinary, such as are seldom recorded. She trained her five children to habits of industry and honesty, making them useful and good citizens. Uniting with great sweetness of disposition, unaffected, frank and winning manners, no one could approach her without loving her.[7]

In the biographical piece on her mother, Sarah Raymond describes her mother, the nature of her work, and the role of women at the time. "The week days of the pioneer woman were full, each household being a factory, and each house-mother was the executive head and managing partner in the business connected therein." Raymond described the work as differentiating the roles of work for men and women, but, of equal importance, each serving the other. Work and entertainment were often mixed. "With all the hard labor, there was happiness, love and truth." What a wonderful description and perhaps a philosophy that kept Sarah Raymond going with her long hours and hard labor as a superintendent knowing she was serving both men and women. She closes the biography with a portrait of her mother as symbol of the pioneer woman. Might this also be a portrait of her as a new pioneer professional woman forging not a new territorial frontier, but a new frontier of women's work and gender roles? Perhaps she is both recording her mother's life work and framing her own.

> Could we have a portrait typifying all the characteristics of a noble, competent pioneer woman, giving to the face strength

The Early Years

and gentleness, showing ability to act as nurse and comforter, the strength to be strong under hardships, a foe to selfishness, wrong and oppression, and full of family love and Christian hope, it might, in a measure, represent what Mrs. Raymond was to her children, and how she lives in the hearts of her descendants.[8]

Sarah Raymond emphasized interesting points in her mother's biography that reveal much about her own views and values and the development of her character. She is clear about the powerful partnership role of women and the importance of their equality and inclusion in any story. She also noted that the work of women, although important, is most often domestic and ordinary and thus left unrecorded. Sarah Raymond challenges this notion by both recording the life work of her mother and of herself and by making her life's work count as a leading female in her day.

Catherine and Jonathan Raymond married in 1831 in Massachusetts. They immigrated with Isaac Whitney to Illinois from Worcester County, Massachusetts, in the autumn of 1834, journeying by stage, the Erie Canal, and a Great Lake schooner. There was no harbor at Chicago, and half a mile from shore the cargo was loaded on scows and towed in. Then a prairie schooner with five yoke of oxen hauled the Raymonds, Mr. Whitney, and their goods to Holderman's Grove. It took five days. They found shelter for the season at the double cabin home of John West Mason, a Yale graduate and settler from Connecticut. In August 1835, Jonathan Raymond and his family settled in Big Grove Township on section 27 W ½ N.E ¼, which was about two-and-a-half miles west of Lisbon on the Chicago and Ottawa stage line, where they built the first frame house on the stage route. Less than a mile away was an encampment of eight hundred Indians. Food was limited. The only vegetable raised was turnips. Those froze that winter. Flour cost fifty dollars per barrel in Ottawa. However, deer, wild hogs, turkeys, and small game were plentiful.[9]

The Raymonds had four sons and one daughter. Sarah Raymond was the middle child with two older brothers and two younger.

Lyman Hamilton Raymond, the oldest son, was a farmer and teacher and filled "offices of trust." George Washington Raymond, the second son, lived in Grundy County, Illinois. He served during the Civil War as captain and later was engaged in farming and trade. Charles Lincoln Raymond, the third son, was described as a man of "rare scholarship and a lawyer of high standing." He left his profession about 1886 to engage in what he considered a more lucrative business. He never married and in later years lived with Sarah in Chicago. Frank Chase Raymond, the youngest son, although studying to be a physician, never practiced but rather engaged in stock raising in Kansas, and later in Texas.[10]

Jonathan Raymond was an active member of the community and built the first Congregational Church outside of Chicago in the northern part of the state. He was interested in music and often directed the music in the church. He was elected sheriff of the county by the new Republican Party in 1856, selected as marshall of the great processions of the political rallies of 1856, and was a noted conductor of the Underground Railroad.[11]

The Lisbon Colony and Big Grove Townships (where the family settled) was known for its radical abolitionists. Many settlers had moved there from Oneida, New York, where there was a training school known as the Oneida Institute of Science and Industry. It combined the search for knowledge with manual labor. This institute was responsible for establishing several colonies in the Midwest in 1835 and 1836. Most of these were in Illinois. One of the purposes of each colony was a plan to buy a township of land at government prices ($1.25 an acre) and sell lots for $75–300 a piece and farmland for $4–10 an acre. The money earned was to be reinvested. Five of the thirty-six sections were intended to build a college with a campus and a college farm. These colonists believed all men ought to have a good education. There is no record why a college was never built in Kendall County, but the records of Knox College in Galesburg, Illinois, show that a large number of students who came to that academy, the college preparatory classes, and college proper were from Kendall County, especially the area around Lisbon. At least three colleges were established by Oneida

Colonies with strong abolitionist ties: Knox College in Illinois, Oberlin College in Ohio, and Grinnell College in Iowa.[12]

Most of the settlers were or became Congregationalists or Presbyterians. The churches were organized under a Plan of Union on an interdenominational basis. Many of the early churches in Kendall County were Congregational. As slavery became a more predominant issue, however, the churches and community split on this issue. The Congregationalists took a firm stand on the abolition of slavery. Those who chose a more moderate path became Presbyterian. Congregational churches were formed as early as 1835: the first at Big Grove and in Bristol in 1836. In 1838, the Lisbon Church was begun, with Newark in 1843, Oswego in 1846, Little Lock in 1853, and Plano in 1858. Members of the Lisbon Congregational Church included Jonathan Raymond and Deacon Isaac Whitney. These early Congregational churches with their Oneida influence were to form the nucleus of the abolitionists, Anti-Slavery Society members, and the heart of the Underground Railroad in Kendall County.[13]

Sarah Raymond Fitzwilliam wrote a six-page article in the *History of Kendall County* about the Underground Railroad. She traces the early roots of slavery, the establishment of the Underground Railroad, and the political, social, and religious leadership in Kendall County. She writes with passion, knowledge, and conviction about the risks and dangers of the Underground Railroad. She writes that her parents were from the state of Massachusetts, and "dyed in the wool" abolitionists and that she was accustomed from early childhood to seeing the "black fugitives off and on at my father's residence until Lincoln's proclamation broke the slave's fetters."[14] She continues, "with my young, girlish eyes I looked on and listened to the strange stories told by these dark brothers in my father's home. These stories have lived with me through these many years." She writes with strong reflection that

> when the complete history of the anti slavery movement in America is written, there will be found in it no chapter so

full of strange and romantic incidents of brave and generous deeds, of moral earnestness in the cause of freedom, and love of liberty for its own sake, as that recounting the work of the Underground Railroad.[15]

In the chapters of her life, she lived episodes in American history reflecting bravery, "generous deeds, moral earnestness in the cause of freedom and love of liberty," to use her words. Her life chapters later included providing equal opportunity for all children to receive an education in a safe and positive learning environment and an opportunity for women to be in professional environments of leadership as teachers and administrators. As a building principal and school superintendent, she fought for the reform of curriculum, discipline, attendance, and employment.

In the closing text on the Underground Railroad she wrote

the old engine of this road has grown rusty from disuse this many a day. The engineers have grown silent, the baying hounds have ceased their pursuit, and Illinois stands glorious in the great galaxy of States with the banner of freedom floating in the breeze for every man, be he black or white.[16]

Perhaps the far-reaching influences of the Underground Railroad inspired a new "Underground Railroad" into the newly emerging frontiers of injustice.

Sarah Raymond witnessed the strength of character and moral conviction to challenge injustice and laws of restriction and emerged as a strong woman forging paths and policies ahead of her time. Despite opposition and public criticism she fought for equality and justice for all. From her family's early abolitionist roots, she learned to be strong and stand up for her beliefs even when they were unpopular. She writes about the past with strong conviction and lived her life with similar passion for the continuing struggle for equal opportunity. It would seem she learned to dare from the examples set by her parents.

Sarah Raymond recalled the significant and powerful sentiment of this characteristic poem of the day recited to her by Rev. John Fry, a graduate of Oberlin College, in the days when antislavery sentiment and Underground Railroads were at a high point:

> I pity the slave mother, careworn and weary,
> Who sighs as she presses her babe to her breast;
> I lament her sad fate, so hopeless and dreary,
> I lament for her woes and her wrongs unredressed.
> O, who can imagine her heart's deep emotion
> As she thinks of her children about to be sold,
> You may picture the bounds of the rock-girdled ocean,
> But the grief of the mother can never be told.[17]

There is no record that any members of the Underground Railroad were ever caught or convicted of crimes in connection with freeing fugitive slaves in Kendall County but their presence was not always well tolerated by fellow citizens as Sarah Raymond Fitzwilliam points out.

> At a 4th of July celebration held at Newark in the 1840s, at which the Rev. H.S. Colton was the impromptu speaker, when he referred to slavery he was jeered and interrupted. James Southworth, himself an abolitionist, told the speaker, for the sake of peace not to agitate the matter. The Reverend Colton then refused to continue.[18]

This text points out an interesting parallel for our own time about the enduring nature of the Underground Railroad.

> They were dissidents, religious nonconformists, who because of moral conviction, actively disobeyed the law by aiding the fugitives, but as fellow citizens they were actively engaged in making the law functional. They were the educated, the intellectual, the middle class who built the schools, built the

churches, laid out the towns, governed the communities and paid the taxes to support them. They were able to dissent without destroying the foundations of their community. They did not compromise their belief in the equality of mankind, but rather taught others through deed and action, that this belief was valid and is essential to the survival of our nation.[19]

This seems to parallel the role of dissent exhibited by Sarah Raymond. She was an educated teacher working within a system but taking on leadership roles uncommon for women in her day and challenging race and gender discrimination of opportunity and pay.

According to Art Thompson, whose great-great grandparents were Jonathan and Catherine Raymond, the Raymond family was friendly with Shabbona. Shabbona was a well-known "peace" chief of the Pottawatomies, who gained prominence during the Blackhawk wars.[20] The Raymonds came to Illinois shortly after the Blackhawk Indian wars and the settlers "lionized" the one chieftain that had remained friendly with the settlers. The story is that Shabbona would ride into the yard and wait on his horse until invited in for dinner. When Shabbona died, George Washington Raymond, Sarah's brother, suggested they provide a natural stone as a marker since Shabbona was a child of nature. The headstone remains in the Morris Cemetery.[21]

We learn through our own experiences and the experiences of others. Sarah Raymond's life is a tale larger than her story alone; it is an opportunity for others to learn, through her, about their own choices and opportunities. J.H. Burnham said "Kendall County has good reason to be proud of one of its early residents, Mrs. Sarah E. (Raymond) Fitzwilliam, whose life history is an example to the younger generation in this and other sections."[22] The context of how she spent her early years paved the way for who she became and the rules she began to break.

She was born in 1842 in what was LaSalle County (now part of Kendall County); there she was raised and began her schooling in

The Early Years

its district schools. *The Pantagraph* reports that

> her first year of school was passed in a little log house, which on Sunday was used as a church. These scenes made such a lasting impression upon her memory that she recalled them in a permanent form by placing in a window of the Congregational church of this city a memorial to her mother, in which was interwoven a picture of this little log house.[23]

She pursued her studies further at an early age of eleven in the Academy at Lisbon, which was near her home. This school was said to have the finest corps of teachers in the West. They were from Vermont and had been selected by Governor Slade of Vermont for their work. By age twelve she was "engaged in the study of French and algebra in addition to the common English branches."[24] The Lisbon Academy was built as a two-story limestone building sometime between the organizational meeting in 1844 and 1849 at a cost of one thousand dollars raised by subscription. When the academy closed sometime after 1884, the building became the Lisbon public grade and high school. The building is still standing and is currently being used as an automotive repair facility. The exterior walls have been covered with stucco so the limestone construction is no longer apparent.[25]

Historian Polly Kaufman writes about the eastern teachers who moved west to teach in frontier schools like the one Sarah experienced and the strong influences of Catharine Beecher and Governor William Slade. Single Protestant women already trained in teaching were recruited and seen as morally suited for the job. These women used teaching as a way to achieve mobility that was acceptable to society. Later Beecher would advocate setting up female seminaries locally for the training of teachers as a cost saving measure, rather than bringing trained eastern teachers west. According to Governor Slade they were "fixed centers of efficient intellectual and religious influence."[26] Kaufman notes that the majority of frontier teachers married and remained active professionally. It is also interesting that "they preserved greater personal independence in their marriages than more provincial antebellum women." They married later in life and had smaller families; several teachers became second wives of men established in the community.[27] These strong pioneer women educators were an early influence on Sarah Raymond, modeling behaviors and choices she would later follow. Raymond also married later in life (as the second wife) after she retired from school work. Many districts, particularly in cities had rules against married teachers continuing to work professionally.

In 1856, Sarah's father was elected sheriff and the family moved to the county seat of Oswego where she attended high school. It was here that she made the acquaintance of Prof. C.D. Wilbur, who became interested in her education advocating the State Normal University in Normal, Illinois, for her future studies. Wilbur was active with the Illinois Natural History Society and taught geology at the University during 1861–62.[28]

However, Sarah Raymond taught school in Kendall County for several terms before attending the Normal University. A piece in the *Journal of the Illinois State Historical Society* suggests she began teaching in Kendall County at the age of sixteen.[29] A review of the 1850 census of Big Grove Township (where she lived with her family) shows she was seven years old and by the 1860 census of Big Grove Township she is listed as age seventeen and a teacher.[30] She is recorded as teaching in four different schools circa 1860.[31] This source is an alphabetical listing of names, schools, dates, and remarks. Sarah E. Raymond is listed four times with the information shown in table 2.1.

In the fall of 1862, at the age of nineteen, she entered the Illinois Normal University at Normal Illinois and completed the course and graduated in 1866. In 1864 the Raymonds moved to Bloomington, in order that their children might complete their education. They remained in the community until they died. Catherine (Sarah's mother) died in Bloomington in 1877. After that, Jonathan (Sarah's father) moved in with Sarah and resided with her until his death in 1884.

The Early Years

Table 2.1 Schools where Sarah Raymond taught

Schools	Date	Remarks
Austin	ca. 1860	Fox Township
Fourth Ward	ca. 1860	School a.k.a. Bushnell
Fowler Institute*	ca. 1860	Big Grove Township
Hollenback	ca. 1860	Teacher in the "Red" school house

*The Fowler Institute was a school in Newark, Illinois, from 1855 until 1880, when it was destroyed by fire. Henry Fowler, a Newark physician, was the founder; his brother Charles became president of Northwestern University. *Chicago Tribune*, March 15, 1964, 24.

After graduating from the ISNU she returned to the Fowler Institute where she remained until 1868 when she left to teach in the Bloomington schools. After five years with the Bloomington schools as a teacher and grammar school principal, she became principal of the high school in 1872 and then was appointed city superintendent of schools in 1873. She held this position until 1892.[32]

In September 1877, The Bloomington *Pantagraph* noted that "Mrs. Raymond, mother of Miss S.E. Raymond, Superintendent of schools, is quite ill at her home in this city, requiring the almost constant attention of her children."[33] Just a few months later, the news of her mother's passing was reported in *The Pantagraph* under the title "Death of Mrs. Catherine H. Raymond."[34] In fact, so significant was her mother's death and so respected was Sarah E. Raymond in the community by that time that "The schools have been ordered to be closed today at noon by Mr. Jacoby, President of the Board of Education out of respect to Miss S.E. Raymond, Superintendent, the daughter of the deceased." Catherine Raymond died of cancer after a lingering illness of five years at the age of sixty-seven just short of her fiftieth wedding anniversary. Jonathan Raymond's death notice was also reported by the *Pantagraph*.[35] He was seventy at the time of his death and had been ill for about one year. He died in the home of his daughter Sarah Raymond, 507 North Mason Street, whom the paper noted "brought him many attentions" in his old age. He was a "devoted Christian and an honest and highly respected man."

The *Pantagraph* obituary for Sarah Raymond noted that her "ancestry was recognized for intellectual strength and aptness" and that "she might be said to have been one of the teachers who were born and not made."[36] Even the closing remarks about her life seem to point to the significance of her early years, her education, and the character of her family, to the teacher leader she becomes. From her early beginnings as a strong student, to her strong role-modeling parents and teachers, and to her strong and powerful family experiences of helping others, Sarah Raymond learned to become a woman who did not back down or shy away from challenge but, rather one who learned to dare and dared to lead others.

During the first half of the nineteenth century women were pursuing educational opportunities both as students and teachers. This was encouraged by several factors; Horace Mann and the common school movement, women like Emma Willard, Catherine Beecher, and Mary Lyon, and the Civil War. This new enthusiasm and interest in education increased the need for schools and teachers. With men called into service for the Civil War, women were afforded an opportunity to enter the labor market as teachers. Middle-class women had few occupational opportunities beyond teaching. Teaching provided an opportunity to contribute to society. Women teachers seemed suitable as the perceived qualities of a teacher were similar to those of mother and homemaker. The number of female teachers continued to increase after the Civil War. Approximately two-thirds of public school teachers were women by 1870. Greater opportunity and equality were available for women in the western part of the United States. As schools increased, so did the number of women who served as teachers and administrators. The early twentieth century saw a steady increase in the number of female teachers and administrators. The upward trend continued until the 1930s when female teachers were a majority but female administrators a minority. The economic crisis and shortage of employment opportunities were hard felt by women as men were perceived as needing the employment more.[37]

How do some women rise to positions of prominence in the field of education? Are there personal qualities that support women toward educational leadership? June Edwards in *Women in American*

Education, 1820–1955 (2002) looks at eight noted women from various areas of education and draws common bonds in their experiences, personalities, and aims. The women studied were: Catharine Esther Beecher, Elizabeth Palmer Peabody, Elizabeth Blackwell, Ellen Swallow Richards, Jane Addams, Maria Montessori, Mary McLeod Bethune, and Helen Pankhurst. Interestingly Sarah Raymond shares in what Edwards identifies as their common experiences. The following qualities are identified by Edwards in her work: great determination to obtain an education and an amazing self-confidence and courage. But where did this will come from, she asks? She suggests with her study of commonalities the following ten elements that empowered these women toward leadership. (1) They came from well-respected families. (2) Birth order or position of responsibility in the family. (3) Admiration and recognition from fathers and supportive mothers who were self-educated and very capable. (4) Although differencing opportunities for boys and girls, if there were boys in the family the girls were seen as having equal value. (5) Managed ways or negotiated a culture of animosity toward women so that their voice could still be heard. (6) Resolute in the goals to improve the lives of all people, seeking justice for those who lacked the power, knowledge, and skills to help themselves. (7) Worked to become professionals in their fields. (8) Willingness to speak out and do the hard work rather than sit back. They were often dismissed by education historians as practitioners rather than leading thinkers of their day. (9) They were supported and encouraged by others in the community both financially and emotionally. (10) They were religious and spiritual. Each of these women followed their career goals at a considerable cost to their personal lives. They had to make choices that men in their day did not have to make. Edwards concludes her introduction with "Reading the biographies and excerpts of these remarkable women, I believe, will give one insight into the times in which they lived and great inspiration for how to live more fully in our own."[38]

CHAPTER THREE

The Illinois State Normal University Years

The women of this institution feel that they have much of which to be proud, much for which to be grateful, as they gather here today, the wards of the great prairie state, the beneficiaries of the famous and timely act of '57... The influence of this institution is not limited to the educational field as narrowly understood. The voice, pen and influence of our women are abroad in the land, in all the various associations and places of honor open to women.

—Sarah Raymond Fitzwilliam,
Jubilee Anniversary Celebration of
the Illinois State Normal University, June 1897

This chapter will address the educational experiences, the teachers, and the role models that influenced Sarah Raymond while she was a student at the Illinois State Normal University (ISNU) from 1862 to 1866. It will also examine Sarah Raymond's relationship with ISNU after she graduated and the impact and influence it had on her career. After completing her early schooling in a variety of settings and teaching in four different schools in Kendall County, Raymond moved about one hundred miles south to central Illinois to continue her study at the Normal University. From her humble beginnings in the small school, to the Lisbon Academy, to Oswego High School, to having her own classroom as a rural school teacher,

Sarah Raymond was well prepared to begin her university studies. The content, pedagogy, and model school training she received at the ISNU would propel her to the top of her field and place her in the annals of scholarship and historical memory.

Brief History of the Illinois State Normal University

The ISNU was founded in 1857 with Charles Hovey as its first principal. The title of president would be introduced later. Principal Hovey served from 1857 to 1861. He was educated at Dartmouth College and was principal and superintendent of schools in Peoria, Illinois, before coming to the ISNU. Sarah Raymond began her studies there in 1862 under the influence of Richard Edwards, the University's second president who served from 1862 to 1876. President Edwards did much to establish the culture and curriculum of the newly emerging university. His task was to improve the faculty and develop a concept of the purpose of the State Normal School.[1] The programs and philosophies he introduced had a lasting impact on Sarah Raymond as a young student, as an emerging teacher, and as a noted educational leader. The University would continue to grow while Raymond was a student there and the changes and expansion of the school made it the best-equipped and most-largely attended Normal School in the United States by the year 1865. In 1867 the Legislature specifically declared the Normal University to be a state institution.[2]

The radical early history of the University is discussed by historian John Freed in his book titled *Educating Illinois*. He makes a compelling argument that ISNU was founded to be a state university of Illinois and functioned that way during the years Raymond attended. It was founded by abolitionists and its earliest plan was to allow all students regardless of race to be admitted. ISNU was established by the legislative Act of 1857. In order to get the act passed, the idea of allowing all students regardless of race to be admitted had to be dropped. It was, however widely discussed in the newspapers of the time. Passing the 1857 Act was the first great accomplishment of the Republican Party after it won the governorship and state superintendency of schools for the first time in the 1856 election. It was the Republicans who voted unanimously for the act. The following are examples of the strong Republican and abolitionist roots of ISNU. The first president of the board was Ninian Edwards, Lincoln's brother-in-law; Jesse Fell, who persuaded Lincoln to write his presidential campaign biography and along with Justice Davis, a major benefactor of ISNU, procured Lincoln's nomination in 1860, was responsible for locating it in Normal; and Lincoln himself was the Board attorney. Jonathan Turner, one of its early founders, was the abolitionist who went to Lincoln in 1862 and urged him to free the slaves; Lincoln pulled out the draft of the preliminary Emancipation Proclamation and told Turner that he would issue it as soon as the Union won a major victory (Antietam). The Raymond family was republican and would have been aware and supportive of the University's early history.

The rank and influence of the ISNU may be estimated from the number of students in attendance. Educational historian Christine Ogren notes that "with the increases in the number of normal schools came increases in enrollment both overall and at individual institutions." She reports that one of the biggest was ISNU in Normal.[3] With this in mind, it is significant that more students were enrolled at Normal in 1863 than had been enrolled in all three of the normal schools of Massachusetts, the birthplace of the normal school institution. In fact, one year after Edwards became president of the ISNU, more students were enrolled in that school than had been enrolled for any one year during the first fifteen years of the Massachusetts normal schools. Five years later the number enrolled at the ISNU had doubled. According to historian Sandra Harmon, students while attending the Normal University lived with local families, faculty members, or rented rooms in boarding houses. School was coeducational with a few exceptions. Men and women took classes together yet Old Main (the only building) had separate entrances and stairways. Campus life was very much coeducational. President Edwards was an outspoken advocate believing women and men were equally qualified to pursue all knowledge.[4] Perhaps it was this early progressive exposure that helped shape Sarah Raymond's

Table 3.1 Student numbers at the ISNU and model school by year

For Year Ending	At the ISNU	At Model School	Total
June 1862	152	133	285
June 1863	205	226	431
June 1864	304	279	583
June 1865	282	411	693
June 1866	270	502	772

ideas about equity of opportunity. The ISNU women graduates had a high employment rate in comparison with some other late-nineteenth-century college graduates, and worked for relatively longer periods of time. Graduates reported holding teaching and administrative positions in all types of schools. Generally, only 7% of students who enrolled graduated; so Sarah Raymond was among an elite group. Sarah Raymond seems to be typical of the extraordinary women at ISNU. There she received the training to help make her a leader in a male-dominated world. At the ISNU men and women students were truly colleagues learning the ways of interaction, oratory, and networking. They took the same classes, were assigned to the same literary societies, lived in the same rooming houses, and took their meals together.

Table 3.1 shows attendance figures during the time Sarah Raymond was a student.[5] The ISNU lists students attending the university and the model school. The model school was a primary and secondary school connected with the university, for the education of students and the training of future teachers.

Admission Requirements

University admission was based on four requirements. Students seeking admission to the university had to apply to the school commissioner of the county in which they resided, and were required:

1. To be if males not less than seventeen, and if females not less than sixteen years of age.

2. To produce a certificate of good moral character signed by a responsible person.
3. To sign a declaration of their intention to devote themselves to school teaching in this state in the form as follows: "I hereby declare my intention to become a teacher in the school of this state; and agree that for three years after leaving the university I will report in writing to the principal thereof, in June and December of each year, where I have been in and what employed."
4. To pass a satisfactory examination, before the proper officers (county school commissioners), in reading, spelling, writing, arithmetic, geography, and elements of English grammar.[6]

The yearly Catalogues of the State Normal University contain interesting information about the members of the Board of Education of the State of Illinois (this was the name of the governing body of the school), faculty, students (both at the university and the model school), course of study, admission, and tuition information. Tuition was free, but books and boarding were not.[7] In the 1862–1863 Catalogue, Sarah Raymond was listed as coming from Lisbon in Kendall County, a member of the junior class and one of 101 ladies and 61 gentlemen. The classes are listed as Junior, Middle, and Senior. In 1863, there were a total of 127 female and 78 male students for a total of 205 students at the university and 226 in the model school. In the 1865 catalogue, Sarah Raymond is listed in the Middle class with a total of 56 classmates (38 ladies and 18 gentlemen). The total university population was 282. In the catalogue ending in 1866, Sarah Raymond is listed in the senior class with a total of 15 students (10 ladies and 5 gentlemen). The university total was 270 and the model school total was 502. The university student numbers declined because few completed the program. Raymond is also listed under pupil teachers-first class. This group was composed of those who had taught four terms in the model school.

The university president, or principal, as they were called at times, was permitted, at his discretion, to "admit to ISNU more than two students from each county, provided the whole number of students did not exceed the aggregate of two from each county and one from each representative district."[8] Many students attended the university but it was a small percentage that graduated. More females attended the university than males, but graduated at a smaller ratio. These trends are discussed in more detail by historian Sandra Harmon in her work on nineteenth-century women at ISNU. She notes "As of 1907, 24,013 students had attended the school since its opening. However, of those only 1,760 or 7.33 percent had graduated from the three-year course of study." And 52 percent of the graduates in the 1860s were women.[9] These trends hold true for Sarah Raymond's experience while at the ISNU. In fact in 1882, President Edwin Hewett reported: "Our course of study is more extended than that of most Normal schools. We insist rigidly on our rule of requiring each one to reach a fixed standard of attainment in any study before he is allowed to pass that study." He explained regrettably the low graduation rate was also due to the fact that many students "are dependent upon their own exertions for means, and, before their course is complete, they are obligated to go out and teach."[10]

Curriculum

The curriculum was a balance between academic and professional course work. *The Catalogue of the State Normal University for the academic year ending June 27, 1862,* Sarah Raymond's first year at ISNU, reveals much of what school requirements were like for Raymond at the time. The university was governed by the Board of Education of the State of Illinois (despite its name it was only governing the university). The Board of Education met approximately four times a year and was comprised of sixteen men from across the state. The catalogue includes information about both the university and the model school. There was a close relationship between the university and the model school. Students taught and observed classes at the model school and faculties of both were expected to model and instruct the latest techniques and approaches to education and teaching. The course of study in 1862 was much prescribed.

In the early curriculum of the ISNU, during Sarah Raymond's time as a student, Homer Hurst reported

> President Edwards specified the following four points as the minimum essentials for prospective elementary teachers. 1. Training in human anatomy and physiology, with adequate attention to gymnastics. 2. A careful and critical review of subjects taught in common school. 3. Instruction in the science and art of teaching (including practice teaching). 4. Courses in the study of English language, literature, and composition.[11]

President Edwards believed that teacher education should include both content and pedagogy. He argued that "a primary duty of the normal schools was to give the student thorough knowledge of the subjects to be taught." A second duty was to "provide professional instruction." He thought that "professional subjects should be the core of the course, with English language and literature the most important of the academic subjects."[12] Under Edwards, the three-year course at the ISNU was so organized that elementary subjects were to be mastered as a prerequisite to the theory courses given during the second and third years.[13]

The second and third years of the plan of study at ISNU included professional subjects such as metaphysics, philosophy, and history of education. It was reported by Burt Loomis that those who took those courses found they had little bearing on the art of teaching. "Didactics, School Laws of Illinois and practice teaching were considered more helpful. Metaphysics and philosophy were included to give fundamental organization to the mind and an explanation of all things."[14]

Mathematics was offered to give "discipline to the mental powers" and physical sciences were given in connection with the Natural History Museum because they furnished excellent solid material on which to "exercise the mental powers." Languages like Latin, Greek, French, and German were offered, but they were considered part of the college preparatory course, and were electives for the ISNU students. In 1874 (eight years after Sarah Raymond attended ISNU) students entered with better preparation and Edwards thus advocated more academic subject matter as a basis for the ISNU curriculum.[15]

Students had instruction in vocal music, penmanship, and drawing among other subjects. Sarah Raymond, class of 1866, described herself as a member of the group of "birds that couldn't sing, and that could never be made to sing," she continued "that the birds" "finally graduated from the pursuit of knowledge under these difficulties, by rising in a body and leaving the hall when the music hour arrived,—no permission being asked or given,—it being tacitly conceded that the pet theory of universal musical training had broken under the strain."[16]

There were three terms in each year of approximately twelve–fifteen weeks each and each term had an assigned course of study. The complete course of study was three years in length. Terms one, four, and seven were fifteen weeks long. Terms two, five, and eight were thirteen weeks long. Terms three, six, and nine were twelve weeks long. Table 3.2 shows the course of study Sarah Raymond followed when she entered ISNU in 1862 according to the catalogue.[17]

According to the *Alumni Register*, there were forty-two teachers who were employed at the ISNU for varying periods of service from 1862 to 1876. Fifteen were teachers in the training department and thirteen were teachers in the model or high school. Most of the remaining fourteen teachers taught several subjects from among the disciplines of mathematics, language, science, history and geography, and drawing.[18] The university community was relatively small during the years Raymond attended and it is conceivable that she would have had contact with all of the faculty and students.

Edwards reportedly believed in practice teaching within an organized program for supervised education and practice. In 1861, a model and practice school with twelve grades had been established, under the supervision of the principal of the high school. In

Table 3.2 Course of study at the ISNU in 1862

Course of Study	First Year			Second Year			Third Year			Terms (weeks)
	1	2	3	4	5	6	7	8	9	
Metaphysics				✓						15
History/methods of ed.		✓				✓	✓		✓	51
Constitution of United States/Illinois								✓		13
School of laws of Illinois									✓	12
English language	✓	✓	✓	✓	✓	✓		✓		93
Arithmetic	✓	✓								28
Algebra			✓							12
Geometry				✓	✓					28
Natural philosophy							✓			15
Book-keeping									✓	12
Geography	✓	✓				✓				40
History				✓	✓					28
Astronomy								✓		13
Chemistry					✓					13
Botany			✓							12
Physiology							✓			15
Zoology									✓	12
Vocal music	All			All			All			28
Writing/drawing	All			All			All			28
Latin language	✓	✓	✓		✓	✓	✓			80
Algebra				✓						15
Higher mathematics								✓	✓	25

1866 and continuing until 1874, the grammar school became a separate department with a separate principal and the high school was a classical preparatory school. President Edwards supervised the pupil teachers in the model school. The faculty observed and critiqued the pupil teachers. All students preparing to become teachers, not only the pupil teachers, were present at the critique sessions to learn and provide feedback.[19]

Edwards accepted no substitute for thoroughness and a professional attitude. Teacher personality, not method, was seen as paramount to the successful teacher. Well-prepared teachers should be ready to serve the community with its changing needs and "finality with regard to education should be avoided." The Normal school should change and expand with the times but avoid "fads and panaceas." Edwards held teachers to a high standard desiring no apologies for, or from them, but rather wanted "enthusiasm for, and devotion to, teaching among his faculty and students at Normal."[20]

Harper, in his historical overview, writes of Edwards as "an inspiring teacher, a splendid leader, and a public speaker of rare power," and calls his belief in a "great purpose" the greatest contribution of Richard Edwards to the cause of education.[21] Teachers should gain in public esteem by meeting the needs of the community, by working toward improving the race, and trying to increase human happiness, Edwards suggested. "Enlightenment for the average citizen in the everyday affairs of life should be in the forefront of the tasks of the competent teacher." Edwards argued that a school for educating teachers had a "field and a function" just as important as any other professional institution. He believed in a system of free public education for students in the primary school to the university. The State Normal School, he asserted, "should be a school of the people." The Normal school concept was necessary because teachers needed scientific training. Normal school trained teachers were efficient and the Normal schools reached the needs of the people, and common school improvement was related to educational progress.[22]

Burt Loomis, reports that Richard Edwards,

> was a master of classroom work, quick of memory, glowing in imagery, fluent in thought and expression. Starting with reading and language, he lifted to a high standard the teaching of the elementary subjects in the state. Students were put through prolonged practice to acquire fluency in reading. This became a recognized characteristic of the teachers trained in this school. This thoroughness set a new standard for teachers in Illinois schools, where there was much slipshod and careless memorizing of meaningless materials in teaching.[23]

The teaching materials Edwards developed, the curriculum that he oversaw, and the preparation of teachers that he supervised was democratic and meaningful. Male and female students participated in a teacher education process where change was occurring. Teachers training at ISNU would be progressive and open to new ways of doing things.

Historian John Freed writes an excellent comprehensive history of Illinois State University using extensive primary documents and secondary sources, *Educating Illinois* (2009). He develops a progressive portrait of a university choosing to integrate both its model school and main campus much to the outrage of others in the early years of its establishment under the thoughtful leadership of President Richard Edwards.[24]

Model Schools

The model school at ISNU was originally established by an order of the board, passed on August 18, 1857, authorizing President Hovey to employ a principal for the model school, should it be necessary. Miss Mary Brooks, who had been a successful teacher under Hovey when he was a principal in Peoria, was the first principal of the model school. It was opened in the fall with only seven pupils, but the number increased to fifty before the end of the year. When the number had increased to almost eighty, the school was divided into two areas. The primary department remained under the direction of Miss Brooks. The responsibility for the more advanced pupils transferred to Mr. G. Thayer.[25] There was some practice and some demonstration teaching in the model school, but little progress was made under Hovey's administration. President Edwards built the small demonstration school at the ISNU into an organized training department.[26] It is interesting to note that the first principal was a woman, and although Sarah Raymond was not yet connected with the school, she would have known of the school's early history and perhaps was empowered by the early position of Brooks in educational leadership.

The model school was from the very beginning an essential part and complement to the Normal school. There were opinions as to whether the model school should be a "practice school for the novice teachers," or whether it should be for "model and demonstration teaching." Cyrus Pierce and Horace Mann, noted educational theorists of the time, favored the practice school point of view. Nicholas Tillinghast, another educational theorist of the time, favored the model and demonstration teaching point of view. President Edwards reserved judgment regarding these issues. He was experienced with both models as a student and faculty member. This exposure to different Normal schools was common. Educational historian Christine Ogren notes there was "considerable cross-pollination among normal schools" in her book *The American State Normal School* (2005).[27] In Salem, Massachusetts, where Edwards taught before coming to the ISNU, he was exposed to the practice teaching concept. At Bridgewater, Massachusetts, where he was a student and taught, he had been exposed to the model school as a demonstration school concept. In St. Louis, where he had been an administrator, he used the public schools for both practice teaching and demonstration. When Sarah Raymond received her teacher training at the ISNU, she experienced the model school as a practice teaching environment and a model demonstration school. In 1862, Edwards described the practice of the Normal students in the model school as follows:

> At the beginning of each term, such members of the higher classes in the University as are designated by the principal for practice in the model school, have classes assigned to them for the term. Each student so designated has charge of one class in one study, and is therefore employed in teaching one hour in the day, the remaining time being appropriated to other work. For the progress of his class during the term, each pupil teacher is held responsible, the principal and other teachers making from time to time such suggestions as the case seems to demand. As frequently as possible, however, the

class, under its teacher, goes through an exercise before the faculty and body of pupil teachers. This exercise is intended to be a fair sample of an ordinary recitation or if allowed to differ from that, it is in order to illustrate more fully some principle or method considered important. After a reasonable time employed in the exercise, the class is dismissed, and the method and manner of the instructor are fully and freely discussed by all present,—their merits and demerits pointed out, and improvements suggested.[28]

Edwards undertook to build up the model school in such a way that not only demonstration teaching, but supervised practice as well, would be furnished for the ISNU students. Because of the additional expense involved, however, it was three years before his plan was fully in operation. The board of education interpreted the act of the legislature that established ISNU as prohibiting the use of funds that had been appropriated to the University for use in the model school; hence the only money available for building up the model school was the tuition from the pupils.[29]

Under President Hovey, the model school had been one school divided into a primary class and a small higher class. The model school was divided into separate primary and high school departments in January 1862. Charles Childs, a principal from the high school in St. Louis, was asked to become principal of the model school and head of the high school department. The primary department continued to be used for practice and demonstration teaching, while the high school was organized as a preparatory school for the colleges and universities initially without practice and demonstration teaching. This organization had just been attained when Edwards came to ISNU and it was President Edwards who oversaw the school's further development with W.L. Pillsbury as principal. When talking about the objective of the department Edwards said:

> One principal object aimed at in the management of the model school during these years, was the thorough fitting of boys for

the best colleges in the country. This, it was thought, would help to give character to the institution in all of its grades. A high reputation for sound scholarship, it was believed, would induce students to come, and would help to maintain good order among them after they were assembled.[30]

Thoroughness in the high school was soon fairly well attained, but to reach its full purpose required the development of the primary and intermediate grades. Edwards also wanted to make practice and demonstration teaching available in all the grades of the model school for the ISNU students. Under his personal attention he reportedly placed some of the strongest ISNU students in charge of classes in all levels of the model school and closely supervised the departments.

Sarah Raymond, when she later became superintendent of Bloomington schools, also personally oversaw all of her teachers and traveled across the city to supervise them. As a new student at ISNU, she observed the changes and institutional improvements Edwards was making and experienced the programs first hand as a student. By the time of Raymond's graduation from ISNU in 1866, the model school structure was highly developed and formalized. W.L. Pillsbury was the principal of the model school including the high school department. Olive Rider was in charge of the intermediate department. Edith Johnson was in charge of the primary department. With this structure there was organization, demonstration, and practice teaching for each grade. Sarah Raymond gained experience at each level and would be prepared to work in all branches of a school as a teacher and administrator.

When criticism of the quality of the teaching in the model school arose in the fall of 1863, Edwards met it in the following manner:

> The moving purpose in establishing the model school was to furnish to the Normal pupils an opportunity for practice teaching...For some years it had been somewhat freely charged that the teaching imparted in it was of a poor quality.

The ISNU Years

Mere pupil teachers, it was argued, could not, in reason, be expected to do as thorough work as well qualified, regularly employed instructors. The natural effect of that objection would be to discourage parents from patronizing the school. As the readiest way of breaking its force, a number of gentlemen from Bloomington and elsewhere—persons well qualified for the work—were invited to give the school a thorough examination, and to report upon its condition and the character of its teaching. Two days were spent in the rooms, listening to the work, and a report was made which effectually turned the edge of all that criticism.[31]

The examination must have served a good purpose because by the year 1863–1864 the model school was crowded with pupils. Sarah Raymond might have learned a lesson from this openness and willingness to confront and challenge opposition.

Wrightonia Society

While at ISNU, Sarah Raymond was assigned to the Wrightonia society. All first year students and new faculty were divided into two debating societies: the Philadelphian or the Wrightonian. Activities included poetry readings, song, debate, lectures, readings, declarations, and socials both within and between the two societies. Meetings took place every Saturday evening. In the early years faculty served as presidents of the societies and they changed on a regular basis. Later, students would serve as president. Faculty members Thomas Metcalf and John Cook both served as president during Sarah Raymond's time with the society. Raymond would have been exposed to a range of ideas in a coeducational environment as a member and she gained leadership experience. By her last year at ISNU, she was a board member of the Wrightonian society. She was elected on February 5, 1866 to the position of editress.[32] Editress was the term given to a female editor. Women enjoyed greater freedom in Normal than their counterparts at, for example, the Industrial University (later called the University of Illinois) where women were not allowed to join the same literary societies as men, nor was there very large campus-wide participation.[33]

In its early years, positions of leadership in these debate societies were held exclusively by men, but on November 13, 1863, it was unanimously resolved that each and every member of the society should be and is eligible to hold an office. Debates were conducted by men, but women commented, presented, and gave musical performances at society gatherings. In fact, on November 26, 1864, both Miss Raymond and Miss Pike, a classmate, critiqued the debate topic "Resolved that under the existing circumstances, the freedom of the press should be restricted." Men and women, following the same order and arrangement or rules and expectations for seating as during class time, sat on separate sides of the hall. These events were social as well as educational. The Wrightonia record book minutes list all the members and Sarah Raymond is listed as member No. 75 joining on September 12, 1862.[34] Honorary members were also listed and by February 7, 1863, 234 people had been issued an honorary membership including No. 21 Abraham Lincoln, Washington D.C. Weekly, there were interesting topics debated or presented. Topics were both serious and comical in nature and both related to education, society, history, or human interest. The minutes further indicate that on Saturday, April 11, 1863, there was a "select reading by Miss Raymond" and on January 7, 1865, an "essay by Miss Raymond" was presented. On March 24, 1866, there was a "reading of the *Oleastillus*," the society's literary paper, "by Miss Raymond." John Freed speculated on the meaning of Oleastellus, which is the Latin diminutive for a species of Calabrian olive tree, the oleaster. "Since the olive was an attribute of Athena/Minerva, the goddess of wisdom, the name of the paper was a learned classical allusion to wisdom."[35]

Some examples of the debate topics included: "Latin language should be made one of the regular studies of the normal course" (September 12, 1863), and "Henry Clay was a statesman superior to Andrew Jackson" (September 26, 1863). Debates also took some humorous approaches. On March 12, 1864, there was a debate based on the Mother Goose rhyme about the old woman who lived in a shoe, "There was an old woman who lived in a shoe who had so many children she did not know what to do; to some she gave meat, to some she gave bread, and some she whipped soundly and put them to bed." The formal debate topic was "Resolved the principles involved in the forgoing lines are immoral and in all respects detrimental to the best interests of our common humanity."[36]

There were many exchanges of ideas related to education. Raymond was learning from her peers and her professors. This stimulating environment of questioning, interesting discourse, and challenging the societal norms would be excellent training for Raymond and her work as superintendent and her daring vision to see and do things differently. Issues of suffrage, gender, universal schooling were just a few of the topics debated. President Edwards gave a lecture on March 3, 1866, with the title "The teacher may be a man." Students debated and discussed during the closing months of Raymond's time in the society whether the teacher or the minister exerts a greater influence for good on the community.

It is clear that the ISNU experience for Sarah Raymond went beyond the classroom. The teaching at the model school and the debate society activities gave her exposure to larger issues of gender, equality, and opportunity. These experiences also gave her the opportunity to express her ideas and challenge those of others.

Sarah Raymond had an impact on fellow students as well. Helen Rudd, a Normal University classmate, made references in her diary to Raymond's ability as a teacher and her involvement with the Wrightonian Society.[37] Rudd comments about a typical day at the ISNU and her encounter with Raymond.

> Went to school, and the same routine of duties were enacted as usual throughout the day study recitation etc. in the evening went to teachers meeting which was very interesting. Miss Raymond's class was examined, and she was praised highly. Wish that I could have had a class this term, the teachers had a

very interesting discussion about the propriety of raising children to "Yes Maam and No Maam" don't agree with Mr. Metcalf about raising his child not to say it.[38]

Interestingly, Rudd left ISNU for a four-month period before finishing her studies to teach in a rural school and she complains in her diary of the hard work, long hours, overcrowded classroom, range of ability, and pay inequality based on gender. She made twenty-five dollars a month and her male counterpart made forty dollars and had fewer students.[39] This shows the reality of the rural school teacher's work environment and the inequality of pay women teachers experienced. Helen Rudd notes in her diary on May 19, 1866, that she "finished cleaning the house today, after dinner washed up and went to hear Miss Raymond play, she was hoping to hear some of the old Normal pieces."[40] According to the minutes of the Wrightonia Society there was a "social" that day.[41]

Criticism of the Normal

The ISNU during its early years was criticized as a needless expense. Critics argued that teachers did not need special training. "When the falsity of this contention was demonstrated by the superior teaching of those trained in Normal University, the criticism was changed to the accusation that those who attended this institution did not teach."[42] A study was conducted to determine the number of graduates of Normal University who were engaged in the art of teaching, and to determine the success of these men and women in hopes of offsetting the criticism. The first survey for this purpose was sent out in 1866 to "men prominent in education throughout the state." Thirty-eight replied and the results were published. A second survey was made in 1869, the results of which were published in the *Illinois School Report* and referred to by Loomis in his book *The Educational Influence of Richard Edwards* (1932). The survey results showed that there was no basis for this criticism.[43]

In 1875, under the subheading "Do the Graduates and Students of this University Teach?" the *Pantagraph* published a list of 692 names of graduates and past students of ISNU who were engaged in teaching.[44] The article that noted this list probably contained only one-half of the actual number of graduates and former students teaching at that time. The *Pantagraph* asked for corrections to be sent to President Edwards, Normal. On the list of 692 names, 12 were listed as professors and instructors in state universities and normal schools, 10 as county superintendents, and 75 as superintendents of city schools, principals of graded schools, and teachers in high schools. Sarah E. Raymond of McLean County was listed as one of the 75.

Graduation

Sarah E. Raymond graduated on June 27, 1866. It is interesting to note that although the course of study was only three years in length Raymond stayed four years. There is no evidence that she took time off to teach as was very common. Either she did not begin her studies until 1863 or she took additional course work to extend her stay. The graduation ceremony began at nine in the morning with prayer and music followed by more music, orations, and essays by all fifteen students in the graduating class. Topics varied. Some were education and teaching related. Others were content oriented from various subject areas. According to the 1866 ISNU graduation program, Miss Raymond presented an essay "Who has lost by the great rebellion?"[45]

In an address that President Edwards began with, "young ladies and gentlemen, members of the graduation class," he shared several stories and lessons that he wished the audience to remember. He talked of the unending destiny that the teacher is to influence. "The strikes of his chisel are not struck upon marble that will surely crumble in the course of a few centuries. No! He labors upon that which is to last longest." The powerful work of a teacher is the development of character, Edwards argued in his speech. He continued to present the counterargument of "practical men" who feel the nature of teaching is life preparation. "Practical men," Edwards said, encourage teachers to teach skills as preparation for work and life. Edwards, however, strongly claimed that in his opinion education is more about the "intellectual and moral culture." The goal of education "is mind, soul, and its culture." There is no nobler work than teaching, he suggested. "The best results of your work, my friends, will not exhibit itself today. The best results of your labors, they will appear only when the soul blossoms in its ultimate perfections! You sow the seed today, but the harvesting is to be in the ending hereafter!" Further in the speech, Edwards said:

> Surely the teacher without patience is as a mariner without a beacon or compass, as a slave animated by no hope,—as a human soul without radiance or inspiration... As you go forth from us today, then, let this word linger with you, what ever else may be forgotten. Persevere. Let patience have her perfect work. Be not discouraged because you cannot see great results, at once. Endure to the end. By the very nature of things, by reason of the very excellence and grandeur of your work, its results cannot appear at once. The very fact of its doing so would prove its comparative worthlessness. It is no mushroom you are building up. The oak is its strength... Cling then to this noble employment. Do not lightly exchange it for another that may promise greater outward honors or more alluring immediate results. Remember that every year ought to add to your effective force in your profession.[46]

President Edwards read aloud the names of the graduating students at the end of his graduation speech. The commencement concluded with the president's address to the graduates and the class song and benediction. Did Raymond realize the impact Edwards and the Normal school would have on her? Did she internalize the message that teachers must feel and believe in this powerful mission in order to sustain the work that is hard, low paying, and often thankless?

Edwards had a strong evangelical ideology about teaching, which came across in the work that he did and the vision he modeled. Raymond is an excellent example of the teacher leader Edwards was describing. She built a long career as an educator of strong moral and intellectual character. She was patient as she worked building herself as a teacher and administrator and the school district as an accredited model. Her career as an educator was as "strong as an oak" and would not be brought down easily.

This was the culture of the ISNU where Sarah Raymond learned to be a teacher and experienced leadership with vision and moral strength of character. There was a strong emphasis on perseverance, patience, and working hard for a higher standard. There were easier and higher paying jobs than teaching, but none that was more important as Edwards suggested. Sarah Raymond did more than stay in the profession. She embodied the modern teacher Edwards talked about. She taught and led with moral and intellectual passion and patience. Serving the schools and the community for eighteen years (as superintendent), Sarah Raymond was an advocate of equal opportunity for students and teachers and tirelessly worked for the betterment of all.

Teacher Institutes

Two strengths of the program at the Normal were the model school and the teacher institutes—both developed by President Edwards during Sarah Raymond's time as a student. In 1863, during Sarah Raymond's first year at the ISNU, the *Chicago Tribune* reported the remarks of President Edwards about the condition of the University showing that the institution has been "constantly increasing in its usefulness and influences" and praising the values of the model school to the Normal students in training them for teaching. He also expressed pleasure that students from more counties were represented than in earlier years. He hoped that more counties would avail themselves of the full benefits conferred upon them by the state. He posed the questions: "How can the influence of the university be more widely diffused through the state and be brought to bear directly upon the educational progress of each and every county of the state?"[47] His suggestions were teacher institutes.

The *Chicago Tribune* covered the 1863 Illinois State Teachers Association meeting held in Springfield, where the organization's president reviewed its history and suggested that the Illinois State Teachers Association, which was founded in Bloomington in 1853, had led the progress of education in the state. The Free School Law of 1855, the office of the Superintendent of Public Instruction, the *Illinois Teacher* publication, and the Normal University were all the results of the efforts of this association. At that same meeting President Edwards suggested that a state institute be held annually at Normal.[48] Only fifteen teachers attended the institute in September of that year; in August 1864, an institute lasting four weeks attracted one hundred and twenty-eight; from 1867 to 1872 the institutes at Normal had an average attendance of two hundred teachers. The State Teachers Institute in 1871 on the third day had an attendance of three hundred and sixty, which the *Pantagraph* noted was larger than usual for the third day.[49] From 1873 to 1879 the sessions were devoted to the study of science, and attendance declined.[50] These institutes were important in their day and were reported in the local paper *The Daily Pantagraph*. County Institutes were also held for teachers connected with city and rural schools in the county. They were advertised in *The Pantagraph* prior to the event with articles by the county superintendent.[51] Examples include the 1874 Teachers' Institute held at Bloomington High School that was assisted by ISNU faculty Prof. John Cook and Aaron Gove[52] and the McLean County meeting also held at Bloomington High School in 1878.[53] It is interesting to see that the work of professional development of teachers has a long and rich tradition. It is also evident that there is a strong link between the area schools and the Normal University. Sarah Raymond as an alumna would have been aware of the local strengths and expertise available through the institutes.

The ISNU Years 57

Raymond attended many of these institutes and even presided over them. Topics and experiences varied. They all focused on content, pedagogy, and even excursions to places of interest. There were day and evening programs. One example was President Edwards' evening lecture on "Causes of Failure among Teachers"[54] and the invitation from Samuel M. Etter, Bloomington's superintendent of schools, to visit the schoolhouses and coal mines of Bloomington. A total of 26 ladies and 24 gentlemen signified their intention of descending into the mines.[55] *The Catalogue of the Illinois State Teachers Institute held at the Normal University August 1869* shows Sarah Raymond from Bloomington, McLean County, among the members along with a total of 147 women and 144 men.[56] The Catalogue from 1871 again lists Sarah Raymond as a member with 121 other ladies and 94 men giving a total of 215 members.[57]

Sarah Raymond benefited from her continued association with the ISNU and its faculty. As a student she was challenged with content knowledge and mentored in pedagogy, and as a young woman she was exposed to progressive ideas about society, race, and gender in a coeducational environment filled with learning and leadership opportunities. The ISNU prepared her to become a strong person, teacher, educational leader, and role model for future generations. She made a difference in the lives of her students, her colleagues, and the larger community.

The influence of ISNU as Sarah Raymond said "is not limited to the educational field as narrowly understood. The voice, pen and influence of our women are abroad in the land, in all the various associations and places of honor open to women."[58]

Later Associations with the Normal University

Evidence of Raymond's later associations, lasting relationship, and valued status with the ISNU is her invitation to return to campus to speak at major events. The year 1882 marked the quarter centennial celebration of the university and Sarah Raymond was a guest speaker representing the class of 1866. Crowded together in celebration were former students, pioneer educators in the state, and prominent citizens of Normal and Bloomington. Sarah Raymond was the class of 1866 speaker and gave a toast for the occasion.[59]

In June 1897, the ISNU celebrated its fortieth anniversary. Hundreds of alumni as well as the three former presidents General Charles E. Hovey, Dr. Richard Edwards, and Dr. Edwin Hewett were in attendance for the two-day event.[60] Presentation topics celebrated the contributions of the University community and its students. Several women were invited to speak and perform musical selections. Sarah Raymond Fitzwilliam was one of them. The ISNU's school newspaper, *The Vidette,* in its coverage of the event noted that Sarah Raymond Fitzwilliam gave one of the most interesting papers of the morning.[61]

The exercises began on Tuesday evening, June 22, 1897, with four addresses by noted individuals formerly connected with ISNU. These were: "The Early Teachers of the Normal School" by Enoch Gastman of Decatur, "The Early Students of the Normal School" by Captain J. H. Burnham of Bloomington; "The Normal School and Dr. Edwards" by Dr. Charles DeGarmo, president of Swarthmore College; and "The Normal School and Dr. Hewett" by Miss Olive Sattley of Taylorville.

On Wednesday, June 23, 1897, the gathering continued, beginning in the morning at 9:30 a.m. with a platform meeting with addresses by Dr. Edwards, the second president, on "Horace Mann and the Normal School," Dr. Hewitt, the third president, on "Nicholas Tillinghast and the Bridgewater Normal School," General Charles E. Hovey, the first president of ISNU, on "The Beginnings of the Normal School in Illinois," Dr. Thomas J. Burrill, vice-president of the University of Illinois, on "The Normal School in the Early Sixties," Edmund J. James, of the University of Chicago, on "Normal Students in Colleges and in Universities," Hon. S. W. Moulton, of Shelbyville, on "The Normal School in the General Assembly in '57," Mrs. Sarah E. Raymond Fitzwilliam on "The Women of the Normal School," and William Hawley Smith, on "The Normal High School."[62]

It was an impressive list of noted educators all connected with the ISNU over the years and from various capacities. *The Chicago Daily Tribune* reported on the "Anniversary Exercises at Normal" and notes "Mrs. Frank J. Fitzwilliam of Chicago, formerly Miss Sarah E. Raymond of Bloomington," spoke. The article ends with a discussion of the elegant closing event banquet in Normal Hall, which over four hundred guests attended.[63]

The 1907 *Semi-Centennial History of the Illinois Normal University* also has a nice account of the fortieth anniversary events of 1897. It reported that "Sarah Raymond recited an original parody on Holmes' Last Leaf, which was a neat bit of verse." The account reads:

> She then continued in a charming style to discuss the Normal women. Women have a genius for teaching. The modern woman—her active business, her education, her clubs—is a different being from the mild-eyed creature of a century ago. We rejoice that emancipation is at hand. More and more the advance of society is adjusting the mind to the new order of things. Particularly as to woman's position, the change is wonderful. The women of the Normal, 6,000 have received instruction here, 1200 of whom graduated. What must be the influence of such a body? The women of this institution have much to be proud of. The state will gladly give its financial support to such an institution. The tone and character of the community is elevated by the women of the Normal, - their influence is broader than the schoolroom. To individualize would be unjust, but proud are we of the names and fame of our women. Great changes have taken place in pedagogies since 1865, and the Normal is a leader in all these. Proud and grateful to thee, your daughters drink to the health of their alma mater.
>
> We'll drink to her past and future, too,
> With thanks for woman's place with you.

Tho scattered ere the setting sun,
Our home is here, our hearts are one.[64]

It was a significant honor and proof of her status and recognition for Sarah Raymond Fitzwilliam to stand together with other noted educators and deliver a significant topic on women, especially in its day. Sarah Raymond valued her Alma Mater and was, in turn, valued by it. She was able to help and give back to, in return, the very institution that helped her as an emerging professional.

She would help preserve early memories of the Normal with the article and verse she wrote for the *Semi-Centennial History of the Illinois State Normal University* titled "The Old Plank Walk" found in the "The Heroic in Student Life" section. In this article, Sarah Raymond describes the construction of the plank walk between Bloomington and the ISNU campus that she and classmates built. She begins by saying: "The story of the old Normal's glories is in other hands, while I am to tell of one of those prosaic accessories, the old plank walk."[65] Interestingly she notes that she is not telling of the glories but rather the prosaic accessories. However significant the event, it is clear that she was not asked to write the primary story of the ISNU's history. She was asked to write an anecdotal story, but one "upon this the very life and comfort of Normal's representatives largely depended."[66] In the article she sets the stage by joking about the challenge of making this a readable story and then writes the story in verse.

Lads and lasses who would go to Normal U.
All must with measured tread the dirt road pursue.
When came chill November's blasts and deluge poured
On the rich black earth, we cried, a board, a board.

Three maidens, with benevolent spirit blest;
Heeded the cry of those who had been distrest.
Their names should be emblazoned in Normal Hall,
For they gathered the sheckels and issued the call.

To Normal's boys who could use hammer and saw,
To meet on Saturday and observe the law
Laid down by the maidens, to build a plank walk
With two boards to a span, so all must "walk chalk."
Other stipulations involved in the case,
Named by the boys, or they'd not enter the race,
Were "the planks for the walk must be known as green
So when sun-dried, the distance the two towns between

Would be divided in half, to give of time more,
To practice phonetics and hearse love lore"
Cracks, between planks, were by agreement narrow,
So lover's tales their neighbors would not harrow.

Historic the day when the hammer and saw
Completed the plank walk according to law.
A feast fit for the gods was spread out of door,
The workers bid to sample the dishes galore.

The setting sun stretched his celestial rays of light
Across the level landscape; 'twas sober-liveried night
When the valiant workers homeward plodded their way
Triumphant in the hearts of the maidens, in work of the day.
The steep, where fame's proud temple shines, is hard to climb,
Flowers are born to blush unseen for lack of time,
But Pike, Dunn, and Raymond of plank-walk fame,
Are blazoned in glory in history's name.

But how can I now from Pegasus descend
And bring this doggerel to respectable end?
Only on bended knee, before the great Muse,
Pray for forgiveness and past abuse.[67]

Sarah Raymond is one of three "maidens" who made history with her vision, group organization skills, and hard work. Together,

Dunn, Pike, and Raymond improved the lives of students. They were noted in their day but sadly, as the poem suggests, their name are not "emblazoned in Normal Hall."

In the same article, Sarah Raymond describes the Bloomington–Normal community and the ISNU campus in the earlier years and attempts to capture the pioneer spirit of the early students and their project to connect the two towns with the old plank walk. No public transportation existed between the two cities. Bloomington in the early 1860s was, according to Sarah Raymond,

> an unpretentious town of possibly ten thousand inhabitants and Normal was its nearest neighbor just two miles away. Normal consisted of the university building, a few boarding houses, and some half dozen private residences. What lay between was prairie, creek, railroad, wagon road, trees, brambles, flowers and all their accompanying associates.

She paints a clear picture of life on the Illinois prairie with her writing. The pioneer sprits, as she described the early girls and boys of the 1860s, helped

> hew the way to comfort and improvement. Nothing daunted them, they feared nothing and each home was a manual training school. Night and day this old walk received innumerable representatives of common leisure. Side by side, walked; thigh to thigh sat scholar, athlete and Bohemian in a guild of fellowship, for better than the dusty ruts of learning,—no fears to beat way, no strife to heal,—the past unsighed for and the future sure,—learning a mutual respect and an appreciation of life which could not be gathered from the contemplation of a cuneiform inscription, or a journey into the wastes of spherical trigonometry.[68]

She also dismissed the need perhaps of agricultural chemistry as a core area of study by remarking "one of our early peculiarities was the possession of ninety acres of land for a model farm, and the existence of the idea that agricultural chemistry, if no more was to be taught in the institution. With the laudable desire to spread a little agricultural knowledge over as large a surface as possible."[69]

Sarah Raymond received an official invitation from President Felmley to attend the February 1914 Founders Day celebrations and the Dedication of Metcalf School. This model school is named after a long-time faculty of teacher education. This building remains on campus today as Moulton Hall. *The Vidette* and the *Alumni Quarterly* had extensive coverage of the events. There was a formal program of speakers and a large group of invited visitors from out of town including alumni and coworkers of Thomas Metcalf were in attendance.[70]

Sarah Raymond's personal invitation and reply indicates her interest and obvious connection to educational leadership at ISNU. The invitation to be at the "old home" as she put it and to have the "privilege of being one to live anew with you the days when Thomas Metcalf led our hosts" pleased her greatly.[71] Her note asks for clarification, however, of the nature of the invitation. Earlier invitations to campus celebrations had her taking the podium to speak, so it is not surprising that she posed the following question in the reply, proposing to talk about the topics of Thomas Metcalf or the Training School or both "I do not understand definitely from your letter the place I am to fill, if any, on that occasion. Do you wish me to 'keep silence'—which is always golden on the part of a woman—or to talk." She is aware of the typical role of women as she poses her question and continues in her closing comments about the speaking invitation "whatever is to be my station I shall be happy," while noting in the text of the letter that she is a "much better listener than talker."[72] From the program, it seems that Sarah Raymond was an invited guest, but not a chosen speaker.

Sarah Raymond was an active member of the ISNU Alumni Association attending events and serving one year as its president. Both the *Alumni Quarterly* and the *Normal School Quarterly* speak of her participation and involvement.[73] She continued her relationship with the Normal School long after graduation. She learned from President Edwards, Thomas Metcalf, other faculty members, and her ISNU classmates more than content and pedagogy; she learned about becoming a female educational leader in the nineteenth century. Historian Sandra Harmon notes in her essay on nineteenth-century ISNU women that

> Sarah Raymond Fitzwilliam was proud of the fact that the "voice, pen and influence of our women are abroad in the land."... Many of the women of the ISNU, both the graduates and those who could only attend for a few terms, did make their influence felt in their communities.[74]

Harmon is referring to Sarah Raymond's 1897 address parts of which were printed in the *Pantagraph*. Sarah Raymond realized the impact ISNU had on her and many others as she addressed the 1897 Jubilee anniversary celebration audience with those powerful words. By that point she had ended her prominent career as a teacher, principal, and city school superintendent. As an educator she influenced many and continued to influence others through her work as an active club women.

In 1918, the *Alumni Quarterly* accurately described the impact Sarah Raymond had with the article commemorating her death.

> In her death the Alumni Association loses one of its most famous members. A list of the positions of responsibility she has held is mentioned...but no publication will ever be able to state the enrichment which came into the lives of thousands because of her work. Her life affords a splendid example of the inability of the statistician to compute the good that is done to society by such an institution as the school which claims her as a member of its alumni. An expert accountant can consult the records of the I.S.N.U. and compute within a few dollars the cost to the state for the four years schooling which this

The ISNU Years 65

institution gave to Miss Raymond, but the accountant does not exist who can tell the worth to society of the results of the inspiration which she received as a student here. The state, through the University, invested a few talents in her education, and she returned that investment a thousand fold.[75]

The Alumni Association did lose one of its most famous members who was enriched by her ISNU experiences and regularly gave back to the Normal University and society.[76]

Historians of education Nancy Hoffman, Christine Ogren, and Sari Knopp Biklen recognize and discuss in their respective books the power of the normal school and the importance it played in the development of female teacher leaders. They suggest the coeducational environment, rich opportunity for experimental teaching, and participation in pre-professional extracurricular activities helped give authentic experiences and voice to women. They wrote about women's rights and debated the role of women in all vocations and taught side by side their male colleagues in the classroom and model schools. Although witnessing the less than equal status of teaching in comparison with other male professions, they saw it to be a source of power and satisfaction. They learned to navigate their new womanhood, cherish their emerging female personality and position of social power, independence, and fellowship.[77]

CHAPTER FOUR

Teacher and Principal of Bloomington Schools

> She was, in my judgment, *par excellence*, the teacher both of pupil and of instructor. She magnified her office and was unwearied in her efforts to promote the cause of higher education in our city.
> —A testimonial quotation about Sarah Raymond by Adlai E. Stevenson, former Vice President of the United States

What did Sarah Raymond do after graduating from ISNU? How did she go from teacher to superintendent? How did she retain the position of superintendent for eighteen years? What impact and influence did she have on the community, the district's schools, teachers, and students? How do we know what we know about her years with the Bloomington school district? What were the educational issues important to the community?

Upon graduation Sarah Raymond took a position in the English Department at the Fowler Institute in Newark, Illinois, in Kendall County where she had previously taught. After two years she returned as a teacher in the Bloomington ward (neighborhood or district) schools where she would spend the rest of her educational career. She worked her way from teacher, to ward school principal to assistant at the high school to high school principal to city superintendent. Her career in Bloomington spanned twenty-four years, from 1868 to 1892.

The sources of information we have about Sarah Raymond's service to the Bloomington schools are both primary and secondary. There is published and unpublished material, both professional and private in nature, by Sarah Raymond. There are yearly District 87 reports that she wrote about the district while she was superintendent. There are numerous newspaper articles and editorials. There are the unpublished minutes from the Bloomington School Board of Education housed in the District 87 Archives. A few school yearbooks and school newspapers exist and can be found in the Bloomington District 87 archives and at the McLean County Museum of History.

Teacher and Ward School Principal

Bloomington grew in population during the long period of Sarah Raymond's involvement with the community and its schools. Census data noted the population in categories of native born, foreign born, and Negro (African Americans during this period were commonly referred to as Negro or Colored). According to census data the 1870 population was 14,590, of whom 10,692 were native born, 3,898 foreign born, and 235 Negro. In 1880 there was a population of 17,180, of whom 13,689 were native born, 3,491 foreign born, and 401 Negro. At the time of her departure, census data for 1890 indicate a population of, 20,464 of whom 16,378 were native born, 4,086 foreign born, and 469 Negro.[1]

Sarah Raymond joined the Bloomington schools in 1868 and began teaching in a second primary grade, in what was known as the "old barn" schoolhouse in the fifth ward. *The Bloomington Leader* noted this was a difficult school with "its wild and prankish pupils" but the character of the young Sarah Raymond was to stay with a project until it had been successfully completed and she did. The admonition to teachers about to undertake the control of certain classes in Bloomington was "fight or fly," the paper reported; and she clearly fought and remained. When she began, her monthly salary was forty dollars but after the first term it was raised to fifty. The old barn school became one of the "brightest and best schools in all this region." By the end of the year she was "rewarded for her faithfulness and efficiency by being elected principal of that school" at a monthly salary of sixty-five dollars. At the close of the year the school moved to a new location. In January 1871, she was placed in charge of Number 5 School with eight assistants and about four hundred pupils at a salary of eighty-five dollars a month. Her work was "very successful and gratifying to herself and the board, and it was one of the most valuable and pleasant of her experiences in this city," the *Leader* reported.[2]

The Number 5 School (Ward 5 School) moved into a building in the sixth Ward (Number 6 school) but continued to be called School Number 5. Miss Raymond remained there until January 1872. She was appointed to open, organize, and take charge of this school and continued until the new Number 5 School was completed. At this time schools were segregated and the district included one colored school (as it was identified).[3]

Raymond reported in the "History of Bloomington Schools" that:

> The colored pupils of the city, had previous to this date, all attended school in a building located on South Madison Street. The distance which some of the colored pupils had to go to attend this school was considered a hardship. Although the laws of Illinois at the time required them to attend the school especially established for them, the colored people of Bloomington decided to make a test case and send the children who lived in the immediate neighborhood of the new building to this school. Several attempts were made and the school authorities maintained the dignity of the law. The children, under order of the superintendent, S.M. Etter, were ejected, whereupon suit of assault and battery was brought by the colored people against the board. The case was carried

from court to court, finally being decided in the Supreme Court in favor of the colored people; and Number 5 school, under my administration, was the first to admit colored children with white children.[4]

The Daily Pantagraph in January 1871 first reported on the incident.

> Considerable excitement exists in the sixth ward school house (new) occasioned by a refusal to admit some colored children who applied for admission to the school. The children continued to press their claims from time to time with a degree of firmness unusual to the race, but so far have been unable to gain admission; but on the other hand have been somewhat violently thrust from the house by a special policeman. The probability is that the difficulty will lead to a legal investigation of the matter, with a view of ascertaining whether colored citizens yet "have rights that white men are bound to respect."[5]

For the next several days *the Pantagraph* covered the events and ran letters from "R" and "Interested Citizen." "R" wrote the issue is "to ascertain if the children of colored citizens have a right, in common with others, in the public schools...We also learn that other suits are being brought in the Circuit Court with the same object in view."[6]

The Pantagraph noted the "difficulty at the Fifth Ward School" and wrote that a suit was filed against the Board of Education. It reported that there was also an indictment against Superintendent of Schools Samuel M. Etter, and special policeman David K. Plumley. (Assault and battery charges were later withdrawn against David Plumley and deemed not worth continuing.[7]) There seemed to be mixed reviews about the situation; some people were in support of the board of education and its efforts to educate the children of color in a separate school, "The effort is utterly uncalled for, and will result in injury to our schools and to the colored children. The Board of Education have done right with the colored people of Bloomington, and they deserve the thanks of the community." Other comments supported equality: "R. says the 'war still wages' in alleging to the efforts now making to secure to all equal advantages in our common or public schools." Equity is the issue.[8] It is not clear who "R" is but perhaps it is Sarah Raymond, principal of the school in question and supporter of integration.

A small side note in the *Pantagraph* indicated the timeliness of this issue and that it was of a larger concern across the state. As the Mayor's Convention proceedings indicate, "discussion of the question of the powers vested in the Boards of Education throughout the State has been the subject of a resolution to the effect that 'the City Council should have equal power with the Board in matters now under the sole control of the Boards of Education.'"[9]

In June 1871, *The Weekly Pantagraph* ran an article: "Colored children in the public schools." The report discussed several cases from the county facing Bloomington Judge Tipton with regard to integrating the schools and admitting colored children to the rights and privileges of the public schools including attending their ward school.[10] Educational historian Robert McCaul includes discussion of two local Illinois cases of integration: *Chase v. Stephenson* (which went to the Illinois Supreme Court) and *Martha Crow, by her next friend v. Board of Education of Bloomington* (which was decided locally and never appealed to the Illinois Supreme Court).[11]

The case involving the students of color in Miss Sarah Raymond's school was *Martha Crow, by her next friend v. Board of Education of Bloomington*. Sarah Raymond challenged the system and the courts by admitting the students and not requiring that they walk to a more distant school. Resolving the issue, June 23, 1871, Judge Tipton decided all the cases of the city and country schools regarding integration together. In Bloomington, he concluded the board did no wrong and provided necessary public buildings for schooling the children of the city and has the power to say where the children should go. "The only remedy for the people is to defeat the Board at the polls and not rely upon the courts. In the country where no special provision has been made, I think colored children have the right to admission to the schools."[12]

Sarah Raymond, who was perhaps "R," helped "wage the war for equality." In 1874, *The Weekly Pantagraph* noted that at the monthly meeting of the board of education an application was made to purchase the colored school property of the city at a price of fifteen hundred dollars since by then the need for a colored school was no longer an issue as schools were no longer divided by race.[13]

She remained at the new Number 5 School as principal until April 1873, when she was chosen first assistant of the high school to serve with the principal Dr. B. P. Marsh. The board elected her to succeed Dr. Marsh when he decided to enter the medical profession during the summer of 1873. "This was one of the very happy years of my professional life," she noted in an article she wrote on the history of the public schools of Bloomington.[14] She served only one year as high school principal before being appointed city superintendent of schools in 1874.

Sarah Raymond saw others take a stand regarding race and education. A product of ISNU, which was a center for progressive ideas of educating all Illinois students together—black and white, it is not surprising that she was empowered to challenge her district administration over integration. The 1855 School Act had provided for free public education only for white children. Some cities in northern Illinois, most notably Chicago but also Bloomington, had gone ahead and made provision for the education of African Americans even though they did not receive state funding. Thus the right of colored children to a free public education was uncertain until the 1870 Constitution guaranteed it. There was already in 1867 a major battle about admitting a colored girl to the combined model/district school in Normal, in which all the major figures in the early history of ISNU participated. John Freed gives a thorough account of this conundrum in his extensive history of the university.[15]

High School Principal

In April 1872, Sarah Raymond was elected to and accepted the first assistantship in the high school. By the close of the school year in June, Miss Raymond was elected to fill the vacancy in the principal position for 1873–74 at a salary of one thousand dollars a year. This well-deserved promotion acted as a spur to higher achievement, and the work at the high school became remarkable for "thoroughness and progress." In 1874, Miss Raymond was reelected at a salary of twelve hundred dollars for the following year 1874–75.[16] In her history of the schools, Sarah Raymond reported that the young people represented the best families of the city and were hearty in their support of any new measures she wished to inaugurate. She had an able corps of assistants, Miss Harriet Dunn and Miss Susan Hale, and the school grew in interest and numbers to about 241. She continued, in her article, the almost exclusive employment of women as teachers, inaugurated during the Civil War, in accordance with the apparent policy of the board of education. No more than two male teachers were employed at any one time from 1857 to 1865 and a total of only twenty-five from 1857 to 1874.[17]

In addition to her teacher and principal duties, Sarah Raymond was active with the Bloomington Teachers Association, attending monthly meetings, and served as secretary from 1870 to 1872. The superintendent served as president. The meetings, which took place in the high school, focused on teaching methods and content knowledge. The Bloomington city teachers discussed concerns and issues and had speakers for professional development. According to an 1871 *Daily Pantagraph* editorial, a visitor enjoyed the opportunity to see Prof. Marsh, the Principal of Bloomington High School, and President Edwards of ISNU deliver interesting presentations. Professor Marsh presented on physiology, which the *Pantagraph* reported "greatly benefited the teachers." President Edwards' lecture was on "Causes of the Failure of Teachers" and challenged parents about what they could do to promote learning at home and support the work of the teachers.[18] According to the Bloomington Teachers Association record book, Normal University faculty, including Dr. Sewall and Prof. DeMotte, joined the discussions as well.[19] On one occasion, Dr. Edwards delivered a paper about free schools. It was noted in the Bloomington Teachers Association record book to be both "valuable and interesting."[20]

Compulsory education was a hot topic in 1871 nationally and at the local level. It was discussed in the *Pantagraph*, among educators at the ISNU, and at professional educational conferences, and organization meetings across the county. It was an issue of state power over the people and some were concerned about the implications for education and who would be dictating the kind of education people were to receive.[21] Another issue receiving some press that year was with regard to the school books. In 1871 a letter to the editor of the *Daily Pantagraph* requested, "Let pupils take their books home."[22] There was some community concern about the fact that books stayed in the schools and were not allowed to go home with the students. The writer referred to the eastern schools where children took their books home for evening review even though the district owned them. In Bloomington the district also owned the books, but students were not allowed to take them home.

Education and local educational issues were becoming community concerns and topics of great interest. The first official high school graduation was held in 1871. Five students received diplomas. *The Pantagraph* covered the day's proceedings and the event itself. There was an invitation to the community to join the celebrations and witness the strength of the school programming. The newspaper reported that the hall was crowded with citizens and Judge Tipton adjourned court. Together with the members of the bar and officers of the court, he attended the exercise. They, together with the board of education, took seats upon the stage.[23]

At about this time, the *Pantagraph* published a piece under the title "The Employment of Lady Teachers in the Bloomington Schools." The article's author was identified only by the letter "S." Might this be the work of Sarah Raymond? The ideas and arguments presented certainly sound reminiscent of her voice and it is not surprising, if written by her, that there is no author's signature. It was a controversial article and could have negatively impacted her position in the district. The ideas presented in the article are ones that Sarah

Raymond held and ones that guided her leadership as teacher, principal, and superintendent. The article was a response to an editorial published in the *Jacksonville Journal* that was critical of women teachers and principals. "S" defends the Bloomington schools, which had been criticized in the editorial, for having female building principals and teachers. "S" ends the letter "Should the editor of the Journal make our schools a visit, it might modify his views of woman's capacity to govern and to impart instruction." The author was clearly passionately connected with the schools and contended that the schools were well managed and well run under female leadership.[24]

While Sarah Raymond was making history serving as one of the nation's first female city high school principals, another woman was making history as well. Georgina Trotter, a partner in a family-run lumber business, was elected as one of the nation's first female school board members. *The Weekly Pantagraph*, in its article "Jacoby, Marsh, Trotter: a Practical Demonstration of Bloomington's Belief in Woman's Suffrage" shared the excitement that surrounded this school board election of Georgina Trotter as the first woman to the Bloomington School Board. "Bloomington has never been a city of radicals and of advanced ideas, yet the placing of a woman in the field in competition with the lords of creation surprised the majority of people."[25] The *Pantagraph* prior to the election reported

> The time has come when public opinion recognizes the fitness of woman to occupy positions of trust and responsibility, especially in the education of youth, and in no city of the Union has a more advanced ground been taken upon this question than in Bloomington. Miss Trotter is a lady who is the possessor of rare business qualifications in addition to many excellencies of character fully appreciated by those who know her, and if elected will fulfill the duties of her office with ability, energy and prudence.[26]

In response to the call to service "Miss Trotter Accepts" ran the *Pantagraph* headline. In her letter of acceptance she wrote that she did not have any particular desire to come before the public and that she was going to decline but "conviction of duty, and a high appreciation of your substantial and practical recognition of the rights of women compel me to accept the invitation which you have generously extended."[27] Twelve years later, the *Pantagraph* covered the April 1886 school board election. It reported "the contest was a curious one" even after over ten years of service to the school board where she was the "leading spirit." The "war seems to concentrate upon Miss Trotter and the chief reason seems to be that she is looked upon as being the particular and personal friend of Miss Raymond." What that exactly means is unclear today, but they were close associates and the public knew that. The article continues with the debate about women in positions of leadership. "Women running the schools is played out!...Women have run the schools well for twelve years." *The Pantagraph* concludes the debate and the article with "Certainly our experience for some years before Miss Raymond—including Gaylord and others—is not inspiring."[28] There seemed to be two issues: that they were close friends and that there was a debate about women running the schools. That same day in the letters to the editor section of the *Pantagraph* someone criticized Miss Trotter and Miss Raymond because they did not have children and the author felt mothers and fathers should be on the school board and running the schools not "bachelors and spinsters." They also refer to Miss Trotter as the "bosom friend of the superintendent."[29] Miss Trotter continued to get reelected and served fifteen years on the school board before retiring because of ill health. She passed away in 1904.

Interestingly, historians Tyack and Hansot suggest a relationship between male school boards and the desire to employ male school superintendents as a way to continue the power balance. "The male superintendents and principals linked the schools to the male dominated power structure of the community and to the male school boards."[30] Perhaps the stage was set for a female superintendent of schools with the presence of a woman on the school board. The traditional sense of power had already been challenged and the community demonstrated openness to new definitions of gender leadership.

Miss Trotter and Miss Raymond became such close friends that Miss Raymond was named executrix of her estate by the terms of Miss Trotter's will.[31] Sarah Raymond was also the executrix of her father's estate, and Trotter and Raymond were noted as witnesses of the signing of his will in 1882.[32] As a provision of the Trotter will, there was to be a Trotter family fountain in Withers Park where it still stands today. Sarah Raymond worked hard and secured Lorado Taft to create it. Much credit was given to Miss Raymond for the vision and completion of the project and was a fine example of the women's lasting bond and commitment. "To Mrs. Sarah Raymond Fitzwilliam much praise is also due for the good taste and constant zeal she has shown carrying out the intent of the Trotter will. Her name will be perpetually linked with the memorial."[33] There are souvenir programs from the memorial fountain dedication in 1911 and numerous newspaper articles chronicling the events.[34]

It was an interesting selection to commission Lorado Taft to do the fountain. He was a noted Chicago artist who taught at the Art Institute and the University of Chicago. He did many public sculptures including work for the Chicago World's Columbian Exposition of 1893. He was from Elmwood, Illinois in Peoria County. Perhaps Sarah Raymond would have had contact with him while she lived in Chicago and was an active member of the Art Institute. Historian Jeanne Weimann captures a little of his story in her book *The Fair Women* (1981). In 1892, Susan B Anthony agreed to sit so that Lorado Taft could do a head for the Woman's Building at the fair. "When word got out some feminists were horrified," Miss Johnson, a female artist, wrote: "I could not stand to think that when Miss Anthony had worked for forty years for women, her bust would be made by a man." Hearing the concerns, Taft wrote Anthony:

> I can put myself in your place sufficiently to appreciate in part the objections which you or your friends may feel toward

having the work done by a man. My only regret is that I am not to be allowed to pay this tribute to one whom I was early taught to honor and revere...Come to think of it, I believe I am provoked after all. Sex is but an accident, and it seems to me that it has no more to do with art than has the artist's complexion or the political party he votes with.[35]

And in the end a compromise was reached and Miss Anthony had two busts made—one by Mr. Taft and the other by Miss Johnson.

Many documents point to the closeness of the friendship and association between Raymond and Trotter. Together they served on the Bloomington library board, and in 1887 a building committee raised seventy-five hundred dollars to build what became the Withers Public Library. It was not nearly enough, but Trotter and Raymond refused to give up and by themselves brought in another twenty thousand dollars.[36] The library emerged as a testament to their commitment and dedication. *The Daily Bulletin* reported in an 1892 article about how when John Trotter, the former mayor of Bloomington and brother of Georgina, had taken ill while traveling abroad in 1889; Trotter and Raymond had gone to New York together to bring him home. He suffered a long illness until his death in 1892.[37] *The Bloomington and Normal City Directory for 1891* lists Miss Raymond as boarding at 801 W. Market Street, the same address as Miss Trotter.[38] This was the first and only year that the directory indicated that they lived together. After an extended trip to Europe in 1913, Sarah Raymond did an interview with the *Pantagraph*. The paper noted that while in Ireland "she visited the birthplace of Miss Georgina Trotter, a well known public spirited woman of the city, who died several years ago." It also noted that she visited with cousins of Miss Trotter and toured a church built by the father of Miss Trotter.[39] The two women shared a strong and committed relationship. They worked side by side to improve the lives of others in the community through their social and professional lives. Historian Sarah Deutsch discusses this new relationship of women in her book *Women and the City* (2000). These "New Women" redefined society and lived together in "Boston Marriages" (female partnerships) claiming new sorts of legitimacy and authority for women.[40] There are no existing letters or correspondence between the friends, so it is difficult to speculate beyond the strong connection that is documented. Together they built the schools of Bloomington as superintendent and school board member and as community activists they built the Withers Library. These two strong women defied limitations based on gender and rose to new heights with the support of each other. They remain together in death, as Raymond is buried in the Trotter Family plot, lifelong friends forever.

Sarah Raymond served only one year as principal of the high school but set the course she followed as superintendent. She supported the students, their families, and the community and on June 13, 1874, delivered a graduation speech to the graduates, their families, and friends.[41] A reception given by Raymond honored the seven graduates of the class of 1874.[42]

In 1911 *The Aegis*, published annually by the senior class of the high school, included a historical look at the school. There was a brief discussion of the size of the graduating classes. In the beginning only two young women graduated in 1864. Small classes of ten–twenty graduated in the 1870s and 1880s, and a larger class of seventy-five in 1909 with a growing school enrollment of over five hundred at the time of the 1911 publication. There is a list of past principals by year and mention of what some did after leaving the position.

It should be noted that there were only three female principals of the high school and Sarah Raymond was the first, serving from 1873 to 1874, followed by Susan Hale, 1874–75, and Harriet Dunn, 1875–84. Some of the early principals went on to various professions. Mr. Bloomfield (1859–61) became a brigadier general in the Civil War. Mr. Hull (1862–64) became president of the Southern Illinois State Normal University. Mr. Marsh (1868–73) became a doctor and practiced in Bloomington. Miss Raymond (1873–74) was "elected city superintendent of schools, a position she filled most efficiently for nineteen years." Mr. Henninger (1885–88) became president of the Western Illinois State Normal School. Mr. Manley (1890–93) had published a number of books—scholarly editions of German classics—that were used widely in the schools of the United States at that time.[43]

Superintendent Gaylord, after a connection of a little less than two years with the schools, resigned in July 1874. Sarah Raymond was elected by the board of education to fill his place. She became superintendent and secretary of the board of education at the yearly salary of fifteen hundred dollars beginning August 4, 1874. As former Vice President, Adlai Stevenson said, she was an excellent teacher of students and instructors and promoter of higher education.[44] She was therefore a natural appointment as replacement superintendent.

The professionally and nationally recognized training of teachers to become school administrators did not emerge until the late nineteenth and early twentieth centuries. Prior early efforts were predominantly related to teacher education and not specifically administration. According to Glass, during the period 1880–1930 teachers, principals, and superintendents became more numerous and were increasingly state licensed and college educated. Many teachers were trained in two-year normal schools and then became administrators. Formal administrative programs grew as communities and school districts expanded.[45] Early in the nineteenth century states recognized the special function of school administration and established the office of state superintendent—New York in 1812 and Maryland, Illinois, Vermont, Louisiana, Pennsylvania, Michigan, and Kentucky before 1840. By the end of the Civil War more than twenty cities had established the office of city superintendent of schools. In 1865 a National Association of School Superintendents was formed, which a few years later became the Department of Superintendence of the NEA (National Education Association).[46] Even with professional organizations and offices, the roles and responsibilities of a superintendent were inconsistent. The relationship with the school board also varied. Some districts had different people for purchasing supplies, overseeing facilities, hiring teachers, and advising curriculum and instruction and little power over the board while other district administrators had immense power and influence and oversaw all areas of school administration.

By 1895, the NEA's Committee of Fifteen formally recommended the position be broken up into business interests and supervision of instruction. The elected school director was to serve as chief executive of the school system and he was to select a superintendent of instruction. As the administrative profession became more defined, the literature about the field began to grow.

CHAPTER FIVE
===

Superintendent of Bloomington Schools

> Education is to train individual humans to work for the betterment of all others of their community and of the world.
> —Sarah Raymond, First Annual Report of Bloomington Schools, 1876–77

Superintendent of Bloomington City Schools

In her history of the Bloomington schools, Raymond writes with passion and excitement about beginning this new position, but also honestly about the challenges that awaited her. She knew the issues and the local papers highlighted them. The community was concerned about her ability as a woman to lead, there was a desire to employ male teachers, and there were issues over curriculum and the mismanagement of funds by the previous superintendent.

This was the first time in the history of this county that a lady had been invited to hold so responsible a post. Active teaching had been so fascinating a profession to me, it was with many questionings in my own heart, that I accepted the very high compliment paid me by the board, but trusted and hoped that all might be as agreeable as my earlier experiences. As I took my seat only a few days before the opening of the school year, I found myself in the midst of the following conditions: out of the 53 teachers employed there were no male teachers, a change had been made in June regarding discontinuing the McGuffey Reader, book keeping errors, a large debt, a disorganized course of study, and aware that many felt a woman could not handle this job."[1]

"A New School Superintendent" the headline read in *The Pantagraph* on July 2, 1874. The article presented a look at the controversial selection by the school board. "Sarah Raymond is to succeed Mr. S. D. Gaylord leaving her position as High School principal where she has been especially efficient and satisfactory." The article notes that "she is a lady who has risen to the front rank of her profession by sheer merit." Her appointment to the new position was seen as "a compliment to her worth but it is also one that will, from its importance, demand the highest order of ability."[2] *The Pantagraph* suggested that this

> step on the part of the Board of Education is a radical innovation on the old system and will of course delight those who like innovations that are adopted in the line of progress and improvement. But it is however, the opinion of many—doubtless the majority—of the people of the city, that this change will not result in the best interest of the schools.[3]

Founded by Jesse Fell, *The Pantagraph* was a Republican-run newspaper. Would Raymond's politics have helped *The Pantagraph*'s comfort level in supporting her for this new position? As the community discussed whether Raymond was fit for the job, able to deal with business matters, and a first rate educator, the *Pantagraph* summarized the duties and qualities needed of the superintendent.

> The superintendent has the care and oversight of all the school property, amounting in value to about $250,000; attends to all repairs, alterations and improvements, and in doing this must of necessity, make the best bargains that can be made

with builders, mechanics, hardware men, carpenters, plasterers, painters, laborers, etc; attends to purchasing coal, wood, books, furniture, and all other school supplies; visits the various school buildings—nine or ten in number; decides difficult cases of discipline that may arise, which can not be disposed of by the principals of schools; conducts teachers' examinations; manages in some degrees the whole corps of teachers, about fifty in number, seeing that they keep their work in prime condition.

In addition to this, the superintendent receives the reports from the various schools and keeps a set of books showing the school statistics, compiled from these reports. Besides all this the superintendent receives and disburses about $60,000 every year, of school funds, and in doing this must of necessity keep a set of books with ledger accounts, and take the necessary receipts and vouchers for the payments made. In addition to this a heavy correspondence has to be kept up, which requires considerable time.

Now who will say that this is not a situation requiring the very highest order of ability? It is not intended to condemn Miss Raymond because she is yet untried. As her friend and admirer, we hope she will be the best and most efficient superintendent the city ever had. But the subject is one of general conversation the city over, and we have adverted to it for that reason.

Those who believe the position can be filled by a woman as well as a man, ridicule the idea that Miss Raymond will fall short of the mark—alleging that she has the business ability, the ambition, the determination and the physical power, to do everything demanded by the position. We trust it may be so, and that the School Board have done right.[4]

The selection of Raymond was made official at the board meeting of July 6, 1874. *The Pantagraph* reported that the board accepted the resignation of Superintendent Gaylord and elected

Miss S. E. Raymond to fill the vacancy at an annual salary of twelve hundred dollars. Miss S. E. Hale was elected to the principalship of the high school at nine hundred dollars annually and Miss Lillibridge to be assistant in the high school at sixty dollars per month.[5]

The Pantagraph ran a correction a few days later in response to this article. Her salary was inaccurately noted and should have read fifteen hundred dollars. There was also another fact the newspaper felt should be known. At the meeting there were only five of seven members present, of whom three voted for her and two for a gentleman. Miss Raymond's apparent majority was a majority of one. It must, however, be remembered that two members of the board, Mr. Folsom and Mr. Wakefield, were out of town, but both had left ballots for Miss Raymond. Had the full board been present, Miss Raymond's majority would have been a majority of five. Two members of the board, Mr. Jacoby and Mr. Marsh, while both personal friends and admirers of Miss Raymond, voted against her as a matter of principle as they preferred that the office superintendency be filled by a man.[6]

The superintendents who proceeded and followed Raymond (all of whom were men) were paid more than she was. S. M. Etter, 1868–72, yearly salary ranging from: first year eighteen hundred dollars to fourth year twenty-five hundred dollars. S. D. Gaylord, 1872–74, with a yearly salary of two thousand dollars. Sarah Raymond, August 5, 1874–August 4, 1892, with a salary ranging from fourteen hundred to eighteen hundred dollars a year. Edwin M. Van Petten, 1892–1901, with a first-year salary of fifteen hundred dollars and a second-year salary of two thousand dollars. The salaries of the high school principal were included for all listed prior to Sarah Raymond but not after. B. P. Marsh served Bloomington High School from 1868 to 1873 with a monthly salary of two hundred dollars. Sarah Raymond is listed next, July 1873–August 1874. Susan E. Hale, who served from 1874 to 1875, and Harriet E. Dunn, from 1875 to 1883, are all listed without salary.[7]

By October, *The Pantagraph* had seemingly put its concerns about a female superintendent aside as it reported positively on the schools after a visit to several: "Our City Schools: Energetic Work and Thorough Discipline, the Old Fashioned Pedagogue and the New Era of Public Education." It also pointed out that the female principals and assistants at the high school have a strong educational background: "Miss S.E. Hale a lady of considerable experience as a teacher and a graduate of an eastern college" and her assistant Miss Hattie E. Dunn, a Normal graduate.[8]

The Pantagraph reported on the "teacher meetings" under the leadership of Miss Raymond. It observed that the teachers were organized into a society (the Bloomington Teachers Association) for the purpose of discussing issues of content and pedagogy with Superintendent Raymond presiding. The regular monthly meetings were recorded in the association record book and often in the newspaper as well. For example, the October 1874 meeting was the first for Raymond as president of the group, so she shared some remarks. She indicated that teachers should secure the cooperation of parents in the training of pupils, that books should be placed in the hands of scholars, that the goal should be to make useful self-reliant citizens, and that meetings should be held in order to give teachers a distinct idea of the kind and amount of work required. However, there was to be no prescribed manner of teaching, provided the teaching was based on "correct" principles. In attendance also that day was Prof. Edwin Hewitt of ISNU who spoke about teaching geography.[9] *The Weekly Pantagraph* went on to say, "Quite a large number of teachers were in attendance" at the November meeting, "and the exercises were more than usually interesting."[10] Tardiness of students and teachers was discussed; there was music and a discussion of teaching various subjects. A speaker was also there from the Ladies' Benevolent Association to talk about establishing an institution for the benefit of others. She hoped a few good women were willing to bestow a little of their wages for the establishment of the home. "Sarah Raymond then addressed the association, saying she was willing to sacrifice anything for such an enterprise. She was willing and anxious to do something to elevate those who were willing to help themselves."

The Pantagraph applauded the work with teachers as

> one of the principal reasons that the city schools of which the city is so proud are as effective as they have been in the past year is because the teachers meet at regular intervals and discuss the various modes of teaching each different branch of knowledge and examine the various defects of the modes which they have been using.

At the time of *the Pantagraph* interview there were fifty-nine teachers in attendance at the Saturday meeting listening to music and fellow educators Miss Raymond, and faculty from ISNU such as Dr. Sewall. The superintendent spoke of the nature of the schools in general and then shared some specific points toward improvement. Teachers asked about ways to get the community more involved in visiting the schools and encouraging the teachers.[11] On December 11, 1876, Professor Hewett lectured on map drawing and Edwards on moral science. The group regretfully noted that they were "losing from their midst our esteemed friend and fellow laborer, President Edwards, so long and favorably known to the workers in the cause of education here and through the state." They would "miss his wise counsel and earnest words and inspiring presence."[12] He left the Normal University that year to accept the pastorate of the First Congregational Church at Princeton, Illinois. In 1885, he became financial agent for Knox College. He was state superintendent of public instruction in Illinois from 1887 to 1891. He served as president of Blackburn University at Carlinville, 1891–93, and retired because of failing health. He died in Bloomington, Illinois, in 1908.[13]

During Raymond's first year as superintendent, she made an effort to keep the community informed about school maintenance issues. In a letter Raymond wrote to the *Pantagraph*, she shared her concerns about why the classrooms were cold. She noted that the weather had been particularly cold and explained in great detail the problems with the furnace.[14] It was clear that she was aware of issues and proactive in seeking a solution. Several people replied in the days following Miss Raymond's letter. One letter was from a former student in the high school and another from a person talking about the specifics of heating air. All were interested and supportive.[15]

During Raymond's first year, the board dealt with several important issues, including increasing the number of graduates from the high school and making more classrooms. The board was concerned that seventy-five students entered and only seven graduated. Trotter made the suggestion to have students do more in the ward schools and, therefore, shorten their course of study at the high school, which they eventually did. The board noted that the district requirements were above and beyond the state requirements.[16] At a different board meeting, Raymond brought up the use of corporal punishment and stated that all teachers had been instructed to report to the principal every case that was deemed necessary. There was a clear attempt to cut back on corporal punishment and the board resolved "that no teacher be permitted to punish a pupil with any inflexible instrument."[17]

At the April 1875 board meeting Sarah Raymond reported concern over inaccurate bookkeeping by her predecessor, Mr. Gaylord. There seemed to be a shortage of $249.60. Others checked the books and the error was not with the numbers provided by Miss Raymond, but from the earlier work of Mr. Gaylord.[18] Sarah Raymond was later reported saying, "Even a crook makes more than a woman" as a dramatic example of the pay inequality between men and women.

The Pantagraph in April 1875 ran a long article analyzing the Bloomington schools and reflecting in particular on the first year with a female superintendent. It said:

> this experiment was one that required a good degree of nerve on the part of the board. The appointment was the first of the kind recorded, in the West at least; and the selection of Miss Sarah E Raymond to this high post of honor was commented

upon generally by the press; Since then, we have heard of the appointment of several ladies to similar positions, among others one to the superintendency of schools of Davenport, Iowa. So Bloomington must be given the credit of making the first departure from the established custom.[19]

A reporter spent several hours over several days visiting all of the ward schools observing the teachers and students. "Everywhere he went, he saw that which convinced him that the service of the City Board and of Miss Raymond and her assistants is well rendered." The reporter noted that there was also a journalist from the Bloomington *Leader* on the same mission. The Raymond administration was very receptive to visitors and must have invited them into the schools. This openness was particularly important for the community in response to the previous superintendent who was accused of taking money. In fact the article cited the following strengths of the schools and the new administration: strong organization of the students and teachers, thoughtful instruction, cost-saving measures, and buildings kept in good condition. Interestingly the article mentioned that with women in positions of leadership the board was saving money. "As Superintendent of schools, Mr. S.M. Etter received a salary of $2,400; Miss Raymond's salary is $1,500. Here is a saving of $900. The salary of Dr. B.P. Marsh as Principal of the High School was $1,800; Miss Hale receives for the same work $1,000. Another item of $800 saved."

Just a few years later, in 1878, the *Pantagraph* published the salaries of teachers at the Normal University as follows: "President Hewett, $3,150; Mr. Metcalf, $1,800 and Mr. Burrington, $1,500 and possibly more, depending on the revenue from the high school, not to exceed $2,000." It is interesting to note that the teachers were paid accordingly, "Miss Paddock, $900, Mrs. Haynie, $900 and Miss Pennell and Miss Miller, $600 each."[20] At the University and at the Bloomington school district women were being paid less than men.

The titles of articles in *The Pantagraph* are provocative at times. For example, "Next Year's Schools. Who Are to Develop the Brains and Do the Spanking in the Ward Schools the Coming Year. The Principal for All the Schools and the High School Corps Not Yet Selected."[21] But the message that seems to come out loud and clear is the support for the work Raymond was doing with the schools.

> The year that has just ended has been one of remarkably successful results, and of almost uninterrupted harmony in the organization of the working force. From the office of the superintendent, down to every department of the system, there has been a continuous, preserving and efficient system of work that has brought about the very best of results, as has been amply demonstrated in the examination with which the terms have ended. Miss Raymond and her corps of teachers are entitled to the greatest praise for the manner in which their arduous labors have been performed.[22]

In July 1875, Sarah Raymond was reelected by the school board to serve as superintendent. *The Pantagraph* covered the special meeting and reported that petitions were presented to the board for consideration. One, signed by 658 voters among whom were the members of the McLean County Bar and leading merchants of the city, asked that Miss Raymond be reelected superintendent. Another petition was from the teachers "showing that if harmony between the superintendent and the teachers is conducive to the welfare of our schools, Miss Raymond's election would certainly have that effect."[23] The petition noted that Sarah Raymond was efficient as superintendent and that her examination of classes has been "thorough and satisfactory." The newspaper reported that a man, who did not have courage enough to face the board himself, called a member out and shared two petitions both requesting a male superintendent. The salary of superintendent was set at fifteen hundred dollars. Once again Raymond received five votes; John Hull, a former Bloomington High School principal received two, and Miss Raymond was declared elected by the chair. Trotter asked to be excused from voting for personal reasons, but changed her mind and voted. An assistant superintendent was engaged at a salary of seventy-five dollars per month for nine months and the board asked Raymond with whom she would like to work. Raymond recommended Miss Harriet E. Dunn, also an ISNU alumni, for the office and the board approved this unanimously. She remained there for several years. The board and superintendent shared in the selection of principals as indicated by the published school board discussions.[24] Interestingly this reappointment was covered by the *Chicago Tribune* as well. The *Tribune* noted that she served the "office with great satisfaction to the people."[25] The paper also noted that "petitions instigated and circulated by one or two meddlesome and chagrinned malcontents had been prepared to overthrow Miss Raymond and put another in her place although she has done more efficient work for much less salary than any of her predecessors for many years."[26]

Just one year after assuming her duties as superintendent, she was proposed as county superintendent. *The Pantagraph* listed a total of four people (the other three were men) being considered for this position. "Whether or not Miss Raymond would accept the nomination remains to be seen, but she is talked about quite strongly."[27] She remained in the position of city superintendent for many more years.

In 1875, *The Pantagraph* gave extensive coverage of the monthly board of education meetings, which included discussion of appointments, textbook adoptions, expenses, and personnel issues.[28] Raymond reported that 2,782 pupils were enrolled and that the high school was crowded, and that more desks were needed. There was a discussion about several students who were suspended because their parents would not buy the textbooks. In that same year Sarah Raymond also reported that during the past month there had been cases of spanking or corporal punishment and the numbers were reported for each school.[29]

The schools grew in importance and interest from year to year, and in 1876 the public schools made a highly praised exhibit for the Centennial Exposition in Philadelphia. Bound volumes of papers from the pupils of all grades on various topics were entered, as well as photographs of the buildings and blackboard work. During the years 1877–78 and 1878–79 a very successful night school was conducted in the high school for those who could not attend the day school with an enrollment of just over one hundred pupils but the program was short lived. And in 1880 the Bloomington High School was placed in the accredited list at the Industrial University (University of Illinois in Urbana/Champaign as it was later called).[30] All these facts demonstrate the progress and growing visibility of the Bloomington schools.

Sarah Raymond notes in the conclusion of the annual report that the debt has been entirely cancelled during the 1881 school year with bonds and coupons regularly and promptly met. She closes

> With the many duties and responsibilities incumbent upon the obligations attaching to my position, there have come many and varied courtesies and kindnesses from all, and a debt of gratitude is thereby incurred, which I feel I can only partially liquidate by giving in return to the people, my best and honest effort to train the children of this city, as far as in me lies, to be noble and useful men and women.[31]

Sarah Raymond demonstrated a sincere passion, interest, and commitment to doing the best job she could for the children of the community and was appreciative of the support she received. *The Daily Bulletin*, another one of Bloomington's newspapers, noted in 1881 that the board of education had met and reported that the "schools are progressing finely and in a first class condition in all respects and the attendance is very large."[32]

The general review of the annual report for 1882 indicates that the year was one of the most trying experiences for Sarah Raymond since occupying the office of superintendent. "The almost continuous rain lasting eight months of the year caused much illness, and at many times it was almost impossible to continue the work in some of the departments. Scarlet fever and smallpox were also diseases which found their way into our ranks," Raymond commented, and "were productive of much anxiety and serious illness."[33] Compulsory law for vaccinations was passed by the State of Illinois, indicating this was a larger educational hazard and concern than just for Bloomington. The Bloomington School Board sent home a letter explaining the Illinois State Board of Health decision requiring vaccinations, which helped make this Herculean task more manageable. Raymond also noted that it was the "extreme kindness and thoughtfulness of our patrons that it was carried out to almost completeness."[34]

The close of the 1882 school year was the twenty-fifth anniversary of the Bloomington public school system. The event was celebrated with a program by members of the past and present boards together with the high school alumni. The weather was terrible and so not all were able to join the program. Mr. Jacob Jacoby, who had been president of the school board since 1873, was present that day and opened the meeting. He spoke about the community, the schools, the administration, and its teachers. He suggested that Bloomington "stands proud, ranking in intelligence and moral character with the most favored, and the schools are provided with the modern conveniences as to comfort and health and constitute the pride of our citizens."[35] "Over 3,500 scholars gather new knowledge at those fountains of education, dispensed by our competent and efficient superintendent, diligent and faithful principals and an earnest and indefatigable corps of teachers. The standard of education in our schools defies the competition of the schools of any city in the State."[36] And the schools, too, are "supported and sustained with a liberal hand by an intelligent and law and order loving people," proclaimed Jacoby. Of the thousands, he continued, who have attended public schools, 200 have graduated from the high school, "whose intelligence and moral character are not excelled by those of any other similar institution."[37]

Others spoke that day at the anniversary ceremony and many wrote letters of regret. Raymond included the letters and speeches of the day in her annual report. Raymond ended the report with a conclusion.

> Thus we present to you fellow workers, members of the board of education, and patrons, as full a view as we could gather of the work of the past twenty five years. In reading the addresses pertaining to the instructors and methods of the early times, there is much for reflection and comparison. May we of the present bear our professional duties with as much dignity, earnestness and holiness of purpose as those who have told us the stories of the "Old Time Schools and Methods." In closing the work of the year, I simply call attention, in addition to what I have already presented, to the statistical tables, which, though occupying a brief space, represent much.[38]

The data reveal significant trends and paint a picture of the values of the district. The tables suggest a significant decrease in suspensions in the public schools over the previous ten years. In 1872–73 there had been 353 suspensions with a steady yearly decline to 1881–82 when there were only 20. With a total student enrollment of 3,447 in the school year 1881–82, the average attendance was the highest at 2,451. The number of cases of tardiness was at an all time low in 1881–82 with 1,262 cases during 1874–75. The number of teachers employed in 1882 was 70, the highest on record.[39]

It is interesting to note that many of the expenditures remained the same during the first years of Sarah Raymond's superintendency. As noted in the 1880–81 annual report and then again in the 1882–83 report, the highest salary paid any teacher per month was $87.50 and the lowest amount was $20. The salary paid district school principals per year was $630 and the high school principal was paid $787.50 per year. The salary paid the superintendent and school board secretary, Sarah Raymond, did go up from $1400 yearly in 1881 to $1550.00 in 1883.[40]

The Seventh Annual Report includes the address of Jacob Jacoby, president of the board of education, to the graduates on June 7, 1883. He congratulated all the graduates but lamented the small number of males in the class.

> I congratulate the lady graduates to their large representation in this graduating class, demonstrating a proper appreciation of a higher education, while I regret the extremely small number of men graduates. Is it a lack of interest on the part of our young men in a higher education, or is it the fault of education? This is a question which deserves the serious investigation of educators.[41]

There is no gender breakdown of classes, but the names of graduates are listed for each year and it is clear there is a larger female population. Is the school with its female leadership attracting and retaining a higher proportion of female students relative to other communities? Or is this a national trend? We do see a similarly high proportion of female students at the ISNU during these years as well.

Harmon, in her work on women at the ISNU, documents that women outnumbered men there as well. Over the course of the nineteenth century, academic qualifications continue to improve for all incoming students and greater numbers of incoming students were high school graduates. She notes that "in December 1897, President Cook reported to the governing body, the board of education, that 202 of the 394 students admitted to the school that term were high school or college graduates. Of those graduates, 163 were women." Although Cook is reporting at a later date this pattern is consistent with the noticeable trend experienced at Bloomington High School. Harmon continues "leading Cook to comment: The great disparity between the number of girls and of boys in the graduating classes of our public high schools shows itself here. But few of our young men have taken such a course. The result is that the women represent a higher degree of culture in our entering classes than the men."[42]

American historian Carl Degler draws similar national patterns in his work on women and the family. He notes that not until 1870 was secondary school education for either sex widespread. In 1870 there were only 160 high schools in the whole country. By 1880, however, the number rose to almost 800 and by the close of the century the number jumped to over 6,000. He writes, "From 1870, when the statistics first began to be kept, until the middle of the 20th century, the girls who graduated each year from high school always outnumbered the boys. It is not surprising, therefore, that the Census of 1880 found that the proportion of literacy for young women was actually higher than for young white men."[43] Bloomington High School and the Normal University are consistent with national trends.

In the closing remarks of the 1883 report Sarah Raymond made several important observations about curriculum, teaching, students, and the schools. She noted that the board had decided by unanimous vote to revise and reprint her *Manual of Instruction* issued in December, 1876. The revision and reprinting was seen as necessary for there had been a great demand for copies from nearly every state in the union, and she noted "those in use in our own schools are so nearly worn out, and not an extra copy remains to meet the demand for our course of work."[44] The board also voted to suspend the training department for one year at least and employ only experienced teachers. This was in reference to an innovation of Raymond, a teachers' training department begun in 1877, where students in the last year of their high school work could prepare themselves for teaching positions in the district's elementary schools, after their graduation.[45] Although seemingly successful as it lasted for over five years, the board dissolved it. The training department program idea perhaps originated in vision and training she learned while at the Normal University.

In her 1883 annual report Raymond anticipated an increase in population with the opening of the new C.&A.A. Rail Road shops and noted the city had already grown rapidly during the past year. The school census taken in April 1883 showed a population of

9,761 persons under twenty-one years of age, 7,551 of whom were between the ages of six and twenty-one. She believed there was an imperative need for more schoolrooms, particularly in the northwest part of the city. She discussed maintenance of the schoolrooms and noted that "painting the plastered walls was a perfect success and the calcimining of school rooms was abandoned. The use of hard chalk this year added to the cleanliness and comfort of the school rooms without adding expense."[46] She also discussed the revised course of study followed during the 1882–83 school year to assign more work to the earlier years in an effort to increase retention and graduation as proposed by Trotter.

Things seemed to be going well and proved the wisdom of the board committee. She noted that

> with the fine corps of teachers, which the Board has elected for the coming year, and with good health on the part of all engaged in the work, and with the school rooms in the comfortable and inviting condition in which they will be placed during the summer vacation, there should be no reason why we should not score one of the most successful years ever known in the history of the schools of Bloomington.[47]

Her closing remarks clearly show the open door-open heart policy—Sarah Raymond had with the community, her teachers, and her students, and the dedication and passion with which she worked.

> We invite our patrons to visit us often; encourage your children by seeing them frequently at their schoolroom tasks. As we close our record of work for '83, without possible opportunity to correct any of its errors, or enjoy again to the fullest extent the pleasure of the real past, we desire to heartily thank all for the uniform kindness, courtesy and genuine happiness which they have helped us to realize amidst our numerous and sometimes perplexing duties, so much that was truly delightful.[48]

The Eighth, Ninth, and Tenth Annual Reports for the years 1884, 1885, and 1886 were published together. Her salary as superintendent/board secretary went up in 1883–84 to $1,600 a year and remained at that level for the next few years. The highest salary paid any teacher per month rose to $122.22. The salary paid to the high school principal went up in 1885–86 to $1,100.[49]

Interesting additions to the curriculum were included in this triennial report and students are listed at graduation with their course of study as well. The options were English, English and German, English and Latin, or English, German, and Latin. By the 1890–91 school year the course of study options were English and Latin, English and German, English, and English Commercial.[50] German had been added to the curriculum in 1871 when the board of education citing the large number of Germans in the community and the larger benefits for all to begin learning a language at a young age voted to include the language in the schools.[51]

A new feature of the report was the listing of the names of the teachers employed in the public schools for a given year and the names of the schools. A total of eleven schools (one of which was the high school) were listed for 1886.[52] Teachers were not clearly identified by gender. Ending this report was an In Memoriam section. The death of a teacher and principal, Miss Mary Gardiner was reported. Her career with the district was highlighted, and it was noted that she had been appointed as assistant teacher in Number 1 School in September 1882. "She remained in that position three years, discharging her duties so acceptably that when a vacancy occurred in the principalship of Number 4 School she was promoted to the place of Principal."[53] This is one of many examples of Raymond's personnel practices. As superintendent, she hired, mentored, and promoted teachers to positions of school leadership.

The school year 1886–87 showed an increase for the first time in the lowest per month salary of a teacher, $30, and the amount the district school principals made per year increased to $765, $810, and $900. The pay range might be based on the fact that there were both men and women in the position of principal at that level. The expense per capita of district school pupils, based on tuition and average daily attendance was $13.62 and the expense per capita of high school pupils was $27.24.[54]

For the first time the report referred to an address at the high school commencement by a noted speaker. George Howland, superintendent of schools in Chicago, Illinois, spoke on "The need of an intelligent scholarship."[55] Again, students presented papers and orations on a range of topics from the courses of study.

The report also included the Form of Contract with Teachers. This contract submitted by Superintendent Raymond offers an interesting look at the profession at this time. The contract in typical fashion states the teacher's name, salary, and conditions. A few points of note:

> The board has the right to discharge the teacher at any time when in their judgment the good of the school requires it.
>
> If a teacher resigns they will forfeit two weeks salary, and the person so elected by the acceptance of this appointment agrees to teach the entire year for which elected and should such teacher know or have reason to believe that marriage or any other circumstance will prevent teaching through the year, this election should be at once declined.[56]

According to the 1887–88 annual report, Sarah Raymond's salary was increased from sixteen hundred dollars to eighteen hundred dollars a year for her work as superintendent and secretary of the board. This was also the last year of board service for her longtime friend and the district's first female board of education member, Georgina Trotter.[57] The thirteenth annual report, 1888–89, has for the first time the inclusion of the Raymond school under the names of different schools. It seems that the previously named Stevensonville School located in the Stevensonville neighborhood in the western section of Bloomington was changed to Raymond School.[58] This was the first school in the city to be named after a person. Surprisingly, it was named both during her lifetime and during her active service to the district. The following schools named would reflect presidents and authors. Previous to this, schools carried a number or a neighborhood title. The Raymond School remains to this day the only school in the district to be named after a teacher/administrator.

The revised course of study, 1890–91, was published in the *Sixteenth Annual Report, 1891–1892* and makes reference to the revised *Manual of Instruction* of 1883 for direction and more specifics.[59] These eighteen pages were the most extensive she wrote. It is clear that Raymond wanted to be as thorough and informative as was possible in her final report. She was clearly laying the foundation and presenting a solid course of study. The *Manual of Instruction* was a groundbreaking book by Sarah Raymond. She presented an organized vision for teaching, administration, supervision, curriculum development, and classroom management.

In the June 1891 school board minutes record book, Raymond did an interesting comparison between Bloomington and other cities in Illinois and their high schools. Table 5.1 presents a summary of the results. It is interesting to note both the items she selected to compare and the results. Bloomington falls short in many categories.[60] Of the six communities surveyed, Bloomington was the smallest in population and high school enrollment. Although small, Bloomington still had similar offerings of foreign languages and a literary society. The length of high school study of three–four years and number of courses taught by a teacher of five–six is consistent as well with other schools. It is, however, very noticeable that Bloomington is the lowest paying district of those she surveyed. Her questions about pay to "lady assistants" draws attention to the fact that other districts had female administrators (a point she was criticized for) and because Bloomington schools had a large female contingent the pay was in turn particularly low. It is also interesting to note that many of the schools did not record the total population by gender, but of the three who did, Bloomington had a noticeably higher number of girls than boys attending while Springfield and Decatur show a much more balanced student population.

Table 5.1 Survey of area high schools (1891)

	Springfield	Peoria	Quincy	Decatur	Bloomington	Rockford
Population	24,963	41,024	31,494	23,500	20,048	23,584
# HS	294	402	180	301	138	289
# of girls	180				114	
# of boys	114			117	24	
# of teachers	8	12	6	7	6	9
# of male teachers	2	7	2	3	2	2
Salary of principal (in $)	1,600	2,000	1,500	1,200	1,200	1,600
Salary of first assistant (in $)	1,200	1,700	1,300	1,000	675	800 (lady)
Highest salary paid to lady assistant (in $)	900	1,200	790	785	725	800
Lowest salary paid to lady assistant (in $)	700	650	600	540	540	650
Total budget for tuition (in $)	6,900	13,500	4,860	5,575	4,490	7,050
# of classes taught by principal	4	3 or 4	3	5 or 6	5	3 or 4
# of classes taught by teachers	5	5			5 or 6	
Length of course of study (in years)	4	4	4	4	3–4	4
Is Greek taught?	Yes	Yes	Yes	Yes	No	Yes
Is Latin taught?	Yes	Yes	Yes	Yes	Yes	Yes
Is German taught?	Yes	Yes	Yes	Yes	Yes	Yes
Greatest number of withdrawals	Second year	First year	First year		First year	First year
Literary societies	Yes	Yes	No	Yes	Yes	Yes

In 1892, *The Daily Bulletin* ran a very interesting contest. A trip to Europe was offered to the county's favorite teacher. Nothing of this size had been attempted outside the major cities and the community was skeptical. The contest seemed to be the talk of the county, however, and *The Daily Bulletin* was a household word. Each vote was cast on a ballot cut from the newspaper. Regular updates were posted on the front page, and efforts were made in the newspaper to encourage voting for one's favorite teacher. The three-month long contest in the end drew over a quarter

Superintendent of Bloomington Schools 103

million votes and was open to any teacher occupying any position from superintendent to primary work engaged at public or private schools. During the contest a second prize was added—a trip to Saratoga, New York, for the National Teachers Convention—and the third prize was a gold watch. The race was decided on May 3, 1892. Miss Nellie Fitzgerald received over seventy thousand votes, Miss Katie Fogarty was in second place with over sixty-four thousand votes and Miss Hattie Hayden was in third place with over sixty-one thousand votes. Sarah Raymond, who was superintendent at the time of the contest, was among the names in the long list of favorite teachers and she received thirty-four votes.[61] All three of the winners were from Bloomington and were graduates of Bloomington High School. Nellie Fitzgerald, the paper reported, was a popular principal of the Raymond School in the Stevensonville section of Bloomington. She taught at several schools in the district for a few years before moving to the

> new and handsome Stevensonville School, which was termed Raymond Hall in honor of the superintendent of schools. She has been identified with this model school since and is now the principal of the institution. Her rise in the regard of the board of education and the superintendent has been rapid, and they were quick to recognize her talents and aptitude in the pedagogical line.[62]

As superintendent, Sarah Raymond began writing an annual report of the Bloomington Public Schools in 1877. The District 87 archives have the reports beginning in 1881, with the Fifth Annual Report and continuing to the present. *The Pantagraph* in 1877 printed the full text of the first annual report and noted its significance.

> The report is a remarkably able paper, and is worthy of attentive perusal and preservation. It reflects the greatest credit upon Miss Raymond and her assistants. The management

of the schools has been reduced to a mathematical accuracy, and in the numerical deductions as to expense attendance are found facts of the greatest value and interest.[63]

The annual report is an important document of the district for each year. It is divided into sections and includes information about a variety of areas. It gives the school calendar and board of education members, officers of the board, standing committees, and school committees. A report of the superintendent, summary of statistics for the year broken down by school, comparative statistics for past years, suspensions over the years, expenditures, numbers of students registered in the district, names of pupils admitted to the high school for that year, a table showing numbers admitted to the high school over the years, the graduation program, textbooks and supplemental materials being used, names of graduates of the high school, and points of business from the school board president regarding expenditures and disbursements. The report ends with a conclusion, which is a summary by the superintendent, and In Memoriam, the recognition section of the passing of a member of the school community. In my analysis I will focus primarily on points of interest or change and the remarks by Sarah Raymond as reported in the annual reports.

The first annual report of the superintendent begins with a lengthy historical sketch of the public schools of Bloomington, Illinois, and includes sections slightly different from later years.[64] The sketch included the district's beginnings in 1857 and its representation by Abraham Lincoln, the first high school, first superintendent, and salaries of male and female teachers. Sarah Raymond noted that women were paid between $25 and $40 and men earned between $50 and $60 per month. A full list to date of school board members was included and Sarah Raymond observed that thirty-three different men and one woman have served on the school board since its beginning. Georgina Trotter, the board's first and, during the era of Sarah Raymond, only female member was elected on April 7, 1874. Sarah Raymond was the eleventh and first woman superintendent of the district with a starting salary on August 5,

1874, of $1500. Salaries of her predecessors ranged from $1,000 for full time work to as much as $2,500. The report also includes the list of former Bloomington High School principals and their salaries. Sarah Raymond was the tenth and first woman building principal since it began in 1857 and was appointed in July 1873 with a monthly salary of $111.11. Salaries over the years ranged from $60 to $200 a month. It is interesting to note that in both cases Sarah Raymond received less than her male predecessor and, therefore, it is not surprising that pay discrepancies for men and women would have been an important issue for her. In this brief history she includes a few facts. The study of German was introduced into the public schools in 1872 and "The first school for the education of the colored children of our city was organized in 1860, and was put in charge of by Mrs. Howard at a salary of $25 per month. This school was sustained until the spring of 1874."[65] In the report the high school principal Harriet Dunn indicated that available seating capacity in the high school was 139, yet average attendance was 176. The average number of pupils per teacher was also high at 42, whereas in many cities the average was 35. In her closing comments Sarah Raymond urged parents to help with attendance and tardiness and noted that the recently published rules and regulations and course of study seemed to be taking shape.

In 1878, *The Pantagraph* published a report written by Sarah Raymond under the title "Our schools: a most complete and interesting annual report of the superintendent. Showing much that is of interest to every citizen of Bloomington. A year of hard work crowned by a brilliant commencement."[66] It is a long title for another important piece by Sarah Raymond. She notes that the statistics reveal an increase in efficiency and power. School management is good, classroom supervision is good, and tardiness and suspensions are down. Corporal punishment is also down as a result of teachers using different disciplinary methods. In fact, so significant was the reduction in tardiness, Raymond commented, that there were over 5,000 tardiness incidents several years ago and at the time of writing the reported number was only 1,949—a decrease of between 60 and 70 percent. She noted that the course of study of the Bloomington public schools was eagerly sought by teachers in all parts of the country, and several applications had been received from school boards interested in buying copies for their whole faculty.

Each year *The Pantagraph* highlighted the annual Bloomington High School commencement. The article listed the program, the board president's address, and other observations. The 1878 commencement, it was noted, was the first evening event and "the public appreciated the change by turning out en masse." [67] The paper also gave credit to the success of the entertainment to the efforts of Miss S. E. Raymond, superintendent of schools, and Miss Hattie Dunn, principal of the High School.

In the conclusion of the 1881 annual report, Raymond notes that the year was the fiftieth anniversary of the town's foundation and twenty-fifth year of the public schools. She indicated that the High School Alumni group numbers one hundred and seventy-eight members and was an organization of which the city might and should be justly proud.[68] In the text she discusses the students, the buildings, and the course of study. "Another year's history of our public school work has passed into the great volume of past events. What ever we may have done, well or ill, is now past recall."[69]

She summarizes the strengths of the graduates who left the district schools with "the discipline they have there attained, the love for good reading, the belief in the honest and honorable labor of both head and hands, are everyday telling the story of the power and usefulness of the instruction imparted in the school room."[70] She compared the school to a home and therefore aimed to keep the "buildings clean, well ventilated and in good repair and brightened by flower, picture and bracket, as convenience and taste may dictate." More importantly, she continues in the 1881 conclusions, that it was "home" because of the

> teachers who are in the closest sympathy with the intellectual, moral and social interests of every pupil—by teachers who

have the independence and love for the work and those under their charge, to be able to drop the conventional dignities of the old time instructor, and come near to the individual characteristics of each and every pupil, as does the parent in home rule and discipline.[71]

Her discussion of the moral and intellectual character seems reminiscent of the lessons she learned as a student of Richard Edwards at ISNU.

Raymond noted that the course of study, adopted in 1876, needed some revisions. Since the attendance had improved and grade-appropriate work was being done more consistently, students were progressing quickly through the material. Thus, the addition of more content to the curriculum was being considered. She noted that a nice variety of textbooks and supplemental materials were being used in the classrooms. This extensive selection was important, it was noted, as it built vocabulary, varied thought, and a confidence that students could read in the parlor as well as in the schoolroom. The student became a good sight reader and gained a knowledge and love of good selections and books. They were reading what was considered at the time standard authors as well as "those more fashionable ones".

The Sixteenth Annual Report, 1891–92 was the last report filed by Sarah Raymond at the end of her last year as superintendent. This document is an important one chronicling her years of service to the schools. The comparative tables show the changes during her eighteen years as superintendent. In 1875–76 there had been fifty-seven teachers employed in the district and at the time of her resignation there were seventy-nine teachers under her supervision. Another significant change was in the numbers of suspensions. In 1872–73 there had been three hundred and fifty-three and in the year 1891–92 there were only thirteen.[72] Might this be due to better or more efficient teachers, clearer administrative direction, or a more pleasant school and learning atmosphere?

Teaching According to Sarah Raymond:
Manual of Instruction

Sarah Raymond's *Manual of Instruction* (1883) is a revision of the earlier edition and was an important contribution to schooling and a significant primary document of past practices. This manual attempted to organize, certify, and standardize how teachers and administrators performed their work. It is a well-organized and clearly indexed document of 278 pages. As the title suggests, it has three main sections: rules and regulations, manual of the schools, and course of study for the high school and district schools. There are two smaller sections at the end regarding the schools boundaries, charter, and amendments. The following discussion illustrates significant elements of the document related to curriculum, teaching, and specifically about teaching history.

In her introduction to the course of instruction section of the manual, Raymond addresses the four purposes: (1) to state a true philosophy on the subject, (2) to give directions and suggestions to equip new teachers, (3) to establish a system of study based on subjects rather than on textbooks, and (4) to maintain a uniformity in the course of study prescribed. Raymond's ideas are consistent with her contemporaries of the common school movement and earlier ideas of Thomas Jefferson.[73] Raymond stated, "In the progress of civilization it has come to be very generally believed that the highest welfare of the individual and of the state demands that all the people shall be educated."[74] She continued that the very existence of a republican government depends upon the general intelligence of the people. She saw the teacher playing three roles in this process: agent, co-laborer, and parent. A teacher must be what he would have his pupils become. The important qualities listed are honesty (truthfulness, justice), kindness, firmness, and sympathy. There are three aspects of instruction: (1) as much content as possible, (2) accuracy and precision in the work, and (3) the proper methods of its presentation. There are three important purposes of school work: (1) the acquisition of knowledge, (2) discipline, and (3) gaining a mastery over the instruments of future cultivation. The manual of instruction suggests pedagogical strategies for teachers to attain these goals. The three-step process suggested by Raymond in her manual is still considered valid today: (1) present a brief review of the previous lesson, (2) engage in the work in the lesson of the day, and (3) arrange and explain the advance lesson.[75] Unlike James Loewen[76] who today writes about the fallacies and problems with U.S. history textbooks, Raymond insists her teachers adhere to textbooks and not challenge their content. She noted the two most prevalent errors are: (1) a slavish adherence to the arrangement and language of the book; and (2) a continued criticism of the book, which results in loss of confidence by the pupil. As many teachers do today, she suggested using it as a resource and letting good instruction and method of teaching drive the use of the book.[77]

The Social Sciences According to Sarah Raymond: Issues and Ideas

As specified by the manual of instruction, there were two courses of study at the high school: a four-year language course or a three-year English course. For each program the same number of history/social science courses and terms were required. Only the year students would take them varied. The following is a list of the six required courses from the social sciences and in parenthesis the year for language/English course of study and terms of each: government (first/first year, one term); physical geography (second/first year, one-and-a-half terms); history (second/second year, two terms); philosophy (third/second year, one-and-a-half terms); geology (third/second year, one-and-a-half terms); geography (fourth/third year, one term).[78]

The views of Sarah Raymond were consistent with what is being advocated today with regard to curriculum and instruction. This plan does not look too much different from what is being currently proposed by the professional organizations of the National

Council for the Social Studies (2000). Raymond's proposal to make instruction topical and relevant is common today and the emphasis of both current professional organizations. In fact, she argues with regard to history instruction that it is too "frequently so conducted as to be distasteful, wearisome and well-nigh fruitless."[79] Those complaints are still valid as a result of poor instructional practices. Raymond suggested integrating geography and history so students see a larger picture of the issues. Other ideas include avoiding too many dates, names, and other details being committed to memory. She suggested that the teacher should rather present an analysis to help students understand connections and the larger story. She suggested that reading the textbook as well as historical novels would further cultivate historical imagination. She wrote that much was to be gained by question and answer, illustrating on the black board, grouping facts, by celebrating anniversaries and birthdays of great men, as well as by varying the methods of instruction. This continues to be true today. Effective instructional strategy for social studies, as history educators, Dynneson, Gross, and Berson suggest, must be varied.[80] As Howard Gardner's work on multiple intelligences has confirmed, students learn differently and thus teachers' instructional practices must cater to that. Educators are no longer asking if one is smart but how one is smart. Raymond suggests the aim in all historical study should be to bind things together and that happens in several ways: (1) by using four natural threads—time, place, persons, and relation of cause and effect, (2) by allowing students to take ownership of their learning by making connections between material already covered and new ideas, (3) by allowing students to think divergently and creatively by making imaginary journeys to other areas or historic periods, (4) by discussing material culture as well, (5) by exposing students to a diversity of customs and cultures, and (6) by encouraging a free expression of ideas and opinions.

As Sarah Raymond suggested in her *First Annual Report of Bloomington Schools*, the purpose of education is to help others better help themselves and their communities.[81] She exemplified that mantra with her own life practices. She made the very most of her early educational opportunities, taught for a period, continued her studies at ISNU, taught again, and became a principal and then superintendent of city schools. Education afforded her freedom, independence, and positions of power, in a time when women were not granted the same opportunities as men. It was her education at home and the schools that allowed her to give back to her community and serve for the betterment of others. It was what she learned that taught her to dare and inspired her to lead others.

CHAPTER SIX

The Resignation

> These [women's clubs] are harmful in a way that directly menaces the integrity of our homes and the benign disposition and character of our women's wifehood and motherhood...I believe that it should be boldly declared that the best and safest club for a woman to patronize is her home.
> —President Grover Cleveland

Resignation of Sarah Raymond as Superintendent

Included in Sarah Raymond's last annual report, the *Sixteenth Annual Report*, was a Retrospect. Raymond begins this twelve-page document by "taking the opportunity of saying some things which it would not have been politic for me to say at any previous time, and which I may not find a convenient opportunity of saying at any future time."[1] Here she refutes or responds to the allegations against her and her administration. After six years of experience with the schools, (from second primary through grammar grades and principalships of grammar and high schools) she was considered as a candidate for the office of superintendent by the board of education. This, she offers, was "wholly unsought and not without its considerations" but "the inducements of a fair salary and promise of hearty cooperation led me to accept the call." Under the heading of Teachers, she noted that there were no men among the seventy-seven teachers who were employed at the time she was appointed superintendent in 1874. In July 1892, at the close of her administration, seven of the seventy-eight teachers had taught throughout her incumbency and three were men. And over the course of eighteen years, three hundred and ninety-two different teachers including substitutes had been employed. Of this entire number, eighteen had been men. Thirty-six teachers had been dismissed for inefficiency.[2]

Raymond made it clear that her goal would be to employ good teachers, regardless of gender, and to keep out the "incompetent and heartless" and that low pay was often an obstacle to achieving that end. The almost exclusive employment of women was inaugurated during the Civil War, and afterward, it had been difficult to recruit or retain male teachers at the rate of pay the district could afford, especially because there was no interest in raising taxes to procure "competent male teachers." Raymond preferred to "employ a first class lady teacher rather than a cheap rate male teacher" and to "let the male teachers prove by practical demonstration that they are fitted for principals and supervisors by actual teaching, and not assign them to such important positions upon the consideration of sex alone. I am for good male teachers or I am for good female teachers, but give the positions of honor and trust to those who show the greatest competency."[3]

The resignation of Sarah Raymond drew the attention of the press at home and beyond. Under the heading of *Educational Gossip* in the *Chicago Daily Tribune,* it was reported that "Miss Sara (sic) E. Raymond, who resigned the Superintendency of the Bloomington (Ill.) schools last June, will spend the winter in Boston. Miss Raymond began teaching in the Bloomington schools since 1865, and since 1874 she has been in charge of the Superintendent's office"[4] Her replacement was also discussed in the *Chicago Daily Tribune.* "The Board of Education this evening selected Prof. E. M. Van Petten, late of the Joliet High School as Superintendent of the public schools of this city. Prof. E.N. Brown of Vallegan, Mich. was first chosen, but he had already contracted

The Resignation

as Superintendent of the public schools of Hastings, Neb."[5] *The Pantagraph* also reported "Prof. E.M. Van Petten chosen." He was a Peoria native who had received his BS and MS from Illinois Wesleyan University in 1885.[6]

What Led to This Resignation?

There was intense excitement in the community and extensive coverage in the press in March 1892 of the Illinois Supreme Court decision granting women the right to vote in school elections.[7] *The Weekly Bulletin*, one of Bloomington's several newspapers, in response to the law granting women the privilege of voting for school officers, reported that the women of Bloomington had been called to meet at the Unitarian Church to discuss the duty of making use of this privilege. "Since the women of Bloomington are concerned for the best interests of our public schools we should regard this matter of voting not as a question of woman's rights but of woman's responsibility."[8] Over one hundred women attended a suffrage meeting in the basement of the Unitarian church.[9] The interest went beyond the papers to the pulpit as well. The First Methodist Church sermon was about "the public school."[10]

The 1892 Bloomington School Board election was a contested race with editorials and candidate support letters a frequent sight in the days preceding the April 4, 1892, election.[11] Some candidates called for change in the school administration and others wished to keep things as they were. *The Sunday Bulletin*, in an attempt to remind people of the school election the next day, listed the four candidates in the following order, Thomas, Green, Quackenbush, and Beath, and noted they were all citizens of standing. Two of the four would win seats on the school board. Interestingly, the first two listed were supporters of the present administration and the last two were outspoken opponents. "There is a great deal of street talk as to the candidates favoring or opposing the present administration and that feature will no doubt enter largely into the expression of the voters. The intention of the women to vote and the report of attempts to deter them on the ground of defects in the law will be watched with lively interest."[12]

The Daily Bulletin noted that the

> W.C.T.U. and the ladies of that organization are all out and are understood to be opposed to the present school administration. On the other hand, many ladies who are friendly to the administration but had not intended to vote are on hand and voting in rivalry. But the excitement and novelty of women at the polls is not the only attraction. There is a big fight on and the men have never before voted so early or so numerously at a school election.[13]

The issues were Catholic versus Protestant teachers, male versus female teachers, and female leadership of the schools.

The Woman's Christian Temperance Union (WCTU) played an interesting and challenging role in politics in its day. It was at times a progressive organization and at others conservative. It focused on various issues in different communities and at different times. By the 1870s the organizations founder, Frances Willard, saw women's suffrage not simply as a right but crucially for women reformers as one of the most indispensable weapons in their struggle to bring about change. Women's need for self-determination, financial independence, and active voice were clear messages of Frances Willard over the years.[14] It is surprising that the local Bloomington Normal WCTU, of which Sarah Raymond was an active member, turned its votes of support away from her and for a return to male leadership of the schools.

The number of women voting was 1,241 out of a total of 3,794 persons voting. Mr. Quackenbush received 2,776 votes and Mr. Beath 2,768 to Mr. Thomas's 995, Mr. Green's 970, and Mr. Soattering's 3. The two victorious school board members were said to represent the sentiment of the opposition to the present government in school affairs. An interesting side note is that the *Bloomington Leader* reported that only "five votes have been cast by colored ladies" and the "reason was that the colored women had decided last week to stay at home Monday and do their work and let their white sisters do the voting."[15] *The Daily Bulletin* noted that the "colored women appeared timid and lacked the self possession of their white sisters."[16] Historian Wanda Hendricks addresses voting practices by African Americans in her book *Gender, Race, and Politics in the Midwest* (1998). She noted that "although race women had become sociopolitical agents," "they still yielded to Victorian mores that dictated that as women, they attempt to usurp neither the prominent political role that black males played nor their limited gains." The implication was that the women would not attempt to challenge male leadership.[17] The use of boycotts is also discussed in the text as a way to convey dissatisfaction.[18] It is difficult to understand if there was a reason the African American women chose not to vote during the school board election of 1892. Were they using their power of boycott, showing support for male leadership, intimidated by or siding with the WCTU criticism of the existing school administration or was the reason that they were not able to get away from their work to go to the polls?

The Bloomington Leader in an article titled "Women Against Women" made the argument that "women resent being governed, directed or in any manner controlled by women" and the tides turned with this election as women for the first time were able to vote.[19]

There seems to be no evidence of any wrongdoing by Miss Raymond or her administration. The problem seems simply to have been her gender, not her ability or professional conduct. "The people of Bloomington in an attempt to reform their educational system need not necessarily ruin the official reputation of a woman founded upon twenty years' service that seems now to have been faithful and conscientious, and which must be so considered until it is disproved and discredited by facts."[20] The conflict seemed to go beyond the school board election to issues of Catholic versus Protestant teachers, management of the school board, or a "ring" of power, women's pay, women in school leadership positions (superintendent and principal), and competition between readership of the local Bloomington Newspapers (*The Pantagraph, The Bulletin*, and *The Leader*).[21]

Interestingly, in the article Sarah Raymond wrote on the "History of Bloomington Schools," she paid tribute to the members of the board of education for their support and concern for the highest interests of the schools. She wrote "from 1874 to 1891 the board was constituted of the trusted men and women of the city and nation."[22] It is perhaps intentional that she did not include the year 1892 in her compliments about the board.

The new board of education for the school year 1892–93 proceeded to organize as recorded in the minutes May 2, 1892. Mr. Funk was nominated for school board president and unanimously elected as he was unopposed. Miss S. E. Raymond and Mr. R. B. Porter were placed in nomination for the office of secretary of the board. For the first time in eighteen years since becoming superintendent and at that point assuming duties of secretary there was an election. With three votes to four, Mr. Porter was accordingly declared elected and took his place as secretary. These minutes were recorded by Mr. Porter. The yearly salaries for secretary and treasurer were voted and fixed at three hundred dollars for the secretary and one dollar for treasurer. This was a continuation of the previous year's compensation. "Dr. Dyson offered a resolution that the thanks of the Board be extended to Miss Raymond for the faithful and efficient manner in which she has conducted the business of secretary of the Board for the last eighteen years, and on motion the resolution was adopted."[23] Another change was recorded on June 6, 1892, in the school board minutes. Jacob Jacoby, a twenty-year school board member deemed it desirable due to health concerns to retire from the board. He had been elected nineteen times as board president, but replaced that year by Mr. Funk in the new board organizations.[24]

At the school board meeting on July 5, 1892, Dr. Dyson nominated Miss Raymond as superintendent, Mr. Quackenbush nominated Mr. Brown, and Mr. Burns nominated Mr. Hancock. "A

The Resignation

ballot being taken resulted in Miss Raymond receiving four votes and Mr. Brown receiving three votes and Mr. Hancock receiving no votes. Miss Raymond was accordingly declared elected superintendent." The board minutes continue with a resignation from Miss Raymond.

> Miss Raymond thanked the Board for the courtesy of re-electing her and stated that arrangements made last January during her visit to Boston precluded her from accepting. The situation that it would require most of the month to close up her work and that she resigned now so that the Board might appoint her successor to take charge when it saw fit and that she would prepare a formal written resignation when prepared.[25]

With no break in the minutes, Mr. Beath moved her resignation to be accepted. Mr. Porter moved "that we elect a superintendent to begin his duties Aug 1, 1892."[26] It was agreed that men would be proposed for the position. For the first time in eighteen years a woman would not be superintendent of Bloomington, Illinois, Schools. The school board agreed on fixing the superintendent's salary at fifteen hundred dollars annually. Mr. E. N. Brown was then placed in nomination by Dr. Wunderlich and Mr. L. G. Hancock by Mr. Quackenbush for the position of superintendent. Mr. Brown received five votes and Mr. Hancock two. Dr. Dyson moved that the election be made unanimous and the motion prevailed. During special meetings of the school board held on July 15 and 18, 1892, the board finally agreed, but Mr. Brown was not available for the position. After several ballots they elected Mr. E. M. Van Petten, as superintendent at an annual salary of fifteen hundred dollars to begin August 1, 1892. On motion of Mr. Quackenbush the election of Mr. Van Petten was made unanimous.[27]

The Bloomington Leader had an extensive article about the "Close of a Career." It was highly complimentary of Sarah Raymond and her years of service to the Bloomington schools. It discussed the reelection and resignation following the April school board election and reported that since January 1892 and the death of an uncle in Boston, Miss Raymond has been "making arrangements to sever her connection with schools at the close of the year." It noted that this fact was known to only a few of her intimate friends as to move on without "excitement or disturbing influences and in order to arrange every detail of the work as perfectly as possible for her successor."[28]

The events leading up to the resignation of Sarah E. Raymond were varied and complex. There were many layers to the situation or pieces of the puzzle that led to the change in administration. There were public and private conversations and decisions that led to decisions. In an era of freedom of the press, multiple newspapers, and the public expression of ideas, it should not be a surprise that there was such scrutiny over the administration of the Bloomington schools. With a shifting climate surrounding the board of education, concerns about the lack of male leadership in the schools, new voting opportunities for women, mistrust between Catholics and Protestants, and a powerful female superintendent for eighteen years, time was right for change. Although opposition or questioning authority accompanies every administration, the year 1892 saw more areas of challenge to Raymond's administration of Bloomington schools than previously. An organized movement of opposition developed. The W.C.T.U. challenged the idea of women in school leadership and organized the vote of women (for the first time) in support of school board members who opposed the sitting school board and school leadership.

Interestingly, in 1893, the *Chicago Daily Tribune* reported on the school board election and noted in the headline that "Women Figure Prominently in the City Elections of Bloomington." In the report it reflected back on the 1892 school board election when Bloomington women first voted, noting that over five thousand votes were cast that year. It discussed the conflict between control of the election by two local banks and the "entrance of the America

THE RESIGNATION

Patriotic Association into the affair for the purpose of opposing the possible effort of Catholics to control the elections." Four board members had been endorsed by the APA, although none had solicited it. Interestingly, the *Tribune* reported, "There was a great outpouring of the women, and this ticket was evidently backed." They also noted "The Catholic portion of the community had no distinctive nominees in the field."[29] So even though there was no "threat," one was created. Women were being organized to vote in a block in both the 1892 and 1893 election. By 1893, however, the underground propaganda of the APA movement was identified in the *Tribune*.

The APA movement (American Patriotic Association) or the new Know-Nothingism was founded in the late 1880s in Iowa. This relatively underground movement was an anti-Catholic organization that by 1893 had entered twenty states, focusing primarily on local elections.[30] Historian Donald Kinzer, in his work on the APA, discusses the movement's impact on several school board elections due to their tactics of community organizing against Catholic teachers or school board members or those sympathetic toward Catholics.[31] And in the 1890s in Illinois and other Midwestern states, with the discussion of compulsory school attendance laws, the APA movement grew in response. Although the laws were not directed against one religion "political contests precipitated partisan use of religious divisions."[32] The APA utilized propaganda to attack individuals or groups, focusing particular attention on the press, women, and temperance as women could vote in school elections in many areas and they argued the press was "whiskey dominated."[33] They confronted and challenged those not supportive of their anti-Catholic philosophies. Elections were won and lost by the powerful impact of this underground movement. Sarah Raymond had been an outspoken supporter of good teachers and all students regardless of religion. She hired Catholic and Protestant teachers alike and asked that religion not be talked about in the classroom. Students also had a choice regarding what schools they attended in the community and teachers where they worked; so neither religion nor compulsory education laws was meant to be an issue for the public schools. There were public, Catholic, and Lutheran choices for schooling. However, it clearly became a concern and even impacted Edwards and the state superintendent position as he was seen as favoring compulsory education and Lutheran pastors turned against him out of concern regarding Catholic versus Protestant education. These conflicts were national. Historians Tyack and Hansot engage these same challenges of Republican versus Democrat, Catholic versus Protestant, male versus female, and native born versus immigrant with regard to school leadership in *Managers of Virtue* (1982).[34] They use Ella Flagg Young as a good example of a woman who was able to transcend traditional roles of women in educational leadership. She became superintendent of Chicago schools in 1909 and was a member of the ISNU board. She, like Sarah Raymond, used female social networks, political organizations, personal levels of educational attainment, and strong character and conviction to help her find support in an unsupportive environment for female educational leaders.

Rather than fight with a school board that now did not unanimously support the superintendent nor her policies and rather than defend herself, her teachers, building principals, and programs to the critical community, Sarah E. Raymond, after being reappointed by the newly elected school board, chose to step down after eighteen years of service as superintendent of the Bloomington schools. She cited as her reason the need to help her family in Boston. She did go to Boston for a period, but then returned to Illinois and married and settled down in Chicago. She resided there until her death in 1918.

The writing was on the wall, Sarah E. Raymond realized there would no longer be "hearty cooperation" between her and the school board and rather than fight that, she chose to resign after a four to three vote for her reappointment as superintendent.

Sarah Raymond ended her history of the Bloomington schools with the idea of character as the ultimate object of all true education, for without it no person is qualified to live. She quoted Sir Walter Scott as saying, "Intellect dazzles, but character leads."[35] At ISNU she learned the strength of moral and intellectual virtues and as a teacher, principal, and superintendent she practiced it. She truly learned to dare and dared to lead.

CHAPTER SEVEN

Leading beyond the Schools: Community Involvement in Bloomington, Boston, and Chicago

> Biography is not taxonomy with the specimen to be reclassified according to the latest findings-it is the story of one life as seen by another, with both always growing and changing.
> —Elinor Langer, *Josephine Herbst* (1983) in *The Challenge of Feminist Biography: Writing the Lives of Modern American Women*

"Miss Raymond has been a remarkably busy woman during all these years, not only in educational work, but she has been active in religious, art, literary and benevolent enterprises."[1] She was active with the Methodist church in Bloomington and was assistant superintendent of the First M.E. Sabbbath School, where she worked closely with her future husband F. J. Fitzwilliam, who was also active with the Sunday school program. She was active in literary affairs in Bloomington, Illinois, where she served as president of the library board and worked closely with Georgina Trotter in building it. She was also president of the Illinois Plato club while in Chicago. Historian Steven Rockefeller in his book on John Dewey discusses The Plato Club and Dewey's involvement with it together with Jane Addams of the Hull House.[2]

Likewise, Allen Davis in his book on Jane Addams discusses both The Plato Club and the relationship between John Dewey and Jane Addams. He observes that Hull House was an educational institution and that Jane Addams was known as a progressive in education who was appointed to several school-related committees.[3] Regarding her benevolent work, *the Leader* article suggests Raymond founded the Bloomington Benevolent Society and prompted the organization of the Industrial School and Home of this city. These organizations helped women who were without means or shut off from society. She was active with many educational organizations. She was president and founder of the School Mistresses' club, twice president of the Woman's State Teacher's Association, president of the Central Illinois Teachers Association, and active with the Woman's Educational Association of Illinois Wesleyan University. In 1888, *The Chicago Tribune* ran an article by her, "The Illinois School Mistresses' Club," where she invited "all earnest, thoughtful, progressive teachers" to join them. This and the Woman's State Teachers Association (which she was president of at the time) were newly formed organizations to have "more intimate acquaintance with those of their profession and for closer professional study."[4] The impact of the School Mistresses Club has never been considered, although Ella Flagg Young suggests it had a tremendous influence on her as an emerging school leader. Historian Joan Smith in her book on Ella Flagg Young notes the impact of the School Mistresses Club and that Young served "a long term as its president."[5] Sarah Raymond thought that women needed the same professional support networks that men had long benefited from.

New and innovative pedagogy had a long and rich tradition at ISNU. The Herbartianism movement as it was called had its beginnings at the time Raymond was a student but began to draw national attention by the late 1800s. Even John Dewey was connected with this Pre-Progressive Era movement.[6]

Raymond notes in her greeting to the women teachers of Illinois article that "Illinois is making rapid strides toward a front rank in educational matters. Nearly 66 per cent of the teaching force of the State are women, and they should arouse to the importance and magnitude of their work." She continues with the following comments.

> We hope the teachers throughout the State will appeal to their respective School Boards and Directors for opportunity to attend this meeting. We believe by careful attention to the thoroughly practical work presented on the program the teachers in attendance may gain an inspiration which will materially improve the schools under their charge. Let every town, city, and township in the state be represented. Those who are young in experience are especially urged to attend. Those of extended experience can do the cause of education untold good by their presence and their suggestions.

The Chicago Daily Tribune ran a follow-up article titled "Eight Hundred Schoolmarms" about The Woman's State Teachers' Association meeting held in Bloomington. Attendees represented all parts of the state and other states as well. Illinois State Superintendent Richard Edwards conducted a class drill in reading. Miss Sarah E. Raymond was reelected as president.[7]

Suffrage, education, and the larger condition of woman in society were important missions for Sarah Raymond and in fact she is noted in several places in Elizabeth Cady Stanton's book *History of Woman Suffrage* (1969), in the chapter on Illinois. She was noted as an efficient school director.[8]

In a tribute to Vice President Adlai E. Stevenson at the time of his death, Sarah Raymond Fitzwilliam spoke the following about the man she called her lifelong friend.

> For several years while I was acting in the capacity of superintendent of your city schools, your distinguished citizen, Mr. Stevenson, was a member of the board of education. I have often in the hours of need and uncertainty sought his advice, and never in vain. To his generous sympathy and wise

counsel I attribute much that I was able to accomplish. Mr. Stevenson had a strong penchant for political life and experiences. Tho of a variant political faith with myself I was always gratified when he won, and to all intents and purposes cast my vote on his side. As the presiding officer of the United States senate he wrung from many a rock-ribbed Republican of old New England their sympathy and regard. He was always a lover of the beautiful in art and possessed some notable historic objects of superb material. But neither this refinement of taste nor his daily life lifted him above willing labor and the tenderest sympathy for those who were rude and unlettered.[9]

It is interesting to note the high level of praise she expressed for him and the closeness and mutual respect they shared in spite of different political party affiliations. In 1892, the *Pantagraph* was challenged politically by the election coverage of Adlai E. Stevenson, a Vice Presidential candidate on the Democratic ticket. The *Pantagraph* was in a dilemma as it was a strongly Republican newspaper and hence could not support Stevenson; however, it did not want to hurt him and so announced it was neutral and sat out that election. The paper was challenged again in 1952 when Adlai E. Stevenson II became the Democratic nominee for president of the United States; the *Pantagraph* again sat the election out.[10]

After leaving Bloomington and her position as superintendent, she moved to Boston. According to her biography in the *History of Kendall County* (1914), she spent several years there and in Cambridge. The article mentions "she was active in clubs, society and public activities and became identified with some of Boston's leading literary and social circles which gave her excellent opportunities for a valuable acquaintance in that center of literary and educational influence."[11] Through her work as secretary of the "All Around Club," one of the largest literary clubs of Boston, the biography continues to note that she became acquainted with E. D. Mead, Jane Austin,[12] Mary Livermore, many of the Harvard professors, Julia Ward Howe, Col. T. W. Higginson, and Oliver

Wendell Holmes. She was also reportedly a member of the Boston Branch of the National Folk Lore Society, the Woman's Educational Society, and the American Academy of Political and Social Science of Philadelphia.

The Pantagraph reported on her marriage to Captain F. J. Fitzwilliam, which took place in Boston in June 1896, where she had been living for the last three years. The couple was married at the home of Sarah Raymond's aunt (372 Main Street, Charlestown, Mass.). In attendance were some friends and family members from Bloomington: Mr. and Mrs. F. Y. Hamilton, Miss Mabel Coblentz, and Miss Madeline Funk. Major W. F. Crawford and his wife, the groom's daughter, formerly of Bloomington were also there. According to *The Pantagraph* report, "The couple plan an extended tour of the eastern states and Nova Scotia followed by a trip up the St. Lawrence and around the great lakes to Chicago. They are expected in Chicago in October and will make that their residence."[13] They lived at 4824 Vincennes Avenue in Chicago.

Perhaps Sarah Raymond met many Chicago notables at Hull House during the regular club meetings and gatherings. The Plato Club met every Sunday afternoon to read philosophic essays. Exhibits were held with art on loan from the Art Institute and Lorado Taft, and John Dewey spoke on occasion.[14] Oliver Wendell Holmes called Jane Addams a "big woman who knows at least the facts and gives me more insights into the points of view of the working man and the poor than I had before."[15]

The two had known each other in Bloomington, where he had been a resident since 1866 and a merchant. He had a large dry goods store in the center of Bloomington, which he had founded with his father and brothers. In 1897, *The Pantagraph* reported a long list of former Bloomington residents now living in Chicago including F. J. Fitzwilliam and wife.[16] His marriage with Sarah Raymond was brief because he died at their residence in Chicago in 1899, just three years later. He had been married previously and had several children. For the funeral of his first wife, *The Pantagraph* reported Miss Trotter and Miss Raymond donated flowers together.[17] At the time of his death *The Chicago Tribune* obituary reported that he left behind two sons and two daughters.[18]

While in Chicago, Sarah Raymond continued her active club and social life. Historian Stacy A. Cordery suggests in her article, "Women in Industrializing America," that the industrialization of the Gilded Age recast the ideology of woman's "separate sphere" and shaped the urban experience. Women were 48 percent of the population during the Gilded Era and with near equality in numbers, things were soon to change. Women working outside the home, the women's rights movement, and the emerging club associations led to the emergence of the "New Woman" and the Progressive Era. In clubs women discussed problems, heightened their awareness, and called for change. These separate female institutions boosted women's self-esteem, turned them into parliamentarians, exposed them to a wider world, put them at ease in public places, and provided them with invigorating and supportive circles of friends and coworkers. The "New Woman" of the Progressive Era, educated, informed, and freer of the separate sphere was clearly the legacy of Gilded Age activists.[19] In fact, historian Anne Firor Scott credits club women with "inventing Progressivism" in the 1880s, which is the movement that followed the Gilded Age.[20]

The Chicago Tribune reported Sarah Raymond Fitzwilliam's involvement with the Daughters of the American Revolution.[21] Both the *Journal of the Illinois State Historical Society* and the *Transactions of the Illinois State Historical Society* document her membership and participation.[22] Records of the Hyde Park Travel Club are found at the Chicago History Museum Archives, and they document Mrs. Sarah Raymond Fitzwilliam's participation. With "Mrs. S.E.R. Fitzwilliams we visited Dublin, a city most auspiciously located; and although four miles from the sea, a noble impression is conveyed, as one nears the city by steamer; Trinity College, with its famous library which contains many rare books, and the museum are most interesting." She delivered lectures and helped with symposium.[23] Activities of the Travel Club included art study, travel lectures, and philanthropy. Art was donated to Hyde Park High School, scholarships were provided to the Art Institute School, and during World War I they turned their attention to the war effort, to list just a few of their activities. Membership was about 170 women.[24] The 1914 Annual Meeting Report of the Travel Club details a lecture presented by Mrs. Fitzwilliam on "Egypt and the Nile."[25]

She was reportedly involved with "Travel Club," "Arche Club," and the "Chicago Woman's Club." A published history of the *Arche Club* records in the In Memoriam section reads: "1918 Mrs. S.E.R. Fitzwilliam '98" indicating she died in 1918 and was a member since 1898.[26] The *Arche Club* focused on art and community. This study club was very active with the Art Institute of Chicago. She returned to Bloomington on occasion to visit and share her stories and adventures.

For approximately one year (1912–13) Sarah Raymond traveled extensively and gave a long interview to a *Pantagraph* reporter detailing her experiences. Among her traveling companions were Miss Madeleine Funk (Bloomington), Mrs. Wilkes (Chicago), Dr. Powers, and Dr. Allen (both from Boston). In preparation for her trip Fitzwilliam applied for her passport on November 1, 1912, in Cook County, Illinois. She indicated that she would be abroad for one year. The applicant was required to sign an oath of allegiance and to have a witness. Her traveling companion Addie M. Smith Wilkes, 4727 Greenwood Ave., signed as the witness, The Chicago Historical Society possesses a copy of the 1908 book by Addie M. Smith Wilkes titled *Autobiography of a Book: Being a Short Story of How this Book was Made*. Fitzwilliam completed the following information for Description of Applicant. "Age: 70, Stature: 5 feet 4 ½ inches, Forehead: high, Eyes: blue-grey, Nose: straight, Mouth: medium, Chin: round, Hair: grey, Complexion: medium, Face: oval."[27]

On her trip she traveled the world, met the Pope in Italy, spent Christmas in Dresden, New Year's Eve in Russia, toured ruins in Egypt, the Holy Land, and Greece, and England and Ireland. In Ireland "she visited the birthplace of Miss Georgina Trotter, a well known public spirited woman of this city, who died several years ago. She also visited with Mr. and Mrs. Hatch, cousins of Miss Trotter."[28] She continued to hold a place of interest and admiration in the hearts and minds of Bloomington residents.

In 1917 and 1918, Fitzwilliam donated several hundred objects to the Chicago Art Institute and Library. The objects in the library collection range from photographs and lantern slides to antique texts and guidebooks.[29] The oldest piece in the Fitzwilliam collection is a 1493 copy of *Liber Chronicarum* or the *Nuremberg Chronicle* as it is known in English, which holds extreme value and historical significance.[30] The art she donated was varied and extensive as well. Some appear to be personal family possessions and others works of collected art; for example, there were textiles, prints and drawings, jewelry, and European Decorative Arts. Interestingly an Italian brooch with a stone mosaic was cataloged with a note bought in Florence in 1913. Perhaps much of her extensive international collection of art was collected during her year-long world tour that she described for the November 29, 1913, *Pantagraph* article.

The probate records for Sarah Raymond are housed in the Daley Center Archives in Chicago since she died in that city. Alvia K. Brown was the executor of the estate. Documents indicate her estate (personal property and real estate) held an estimated value of $38,833.07. The funeral costs were $384.82 after which she was transferred to Bloomington on April 10, 1918, for burial. The fare to Bloomington and expenses were included ($51.32). The bill itemizes the cost for her presentation, steel gray casket and trimmings, music, minister, and carriage.

Preparation for her death took place over about a year. In December 1917 she was making large donations to the Art Institute. She contacted the Bloomington Public Library to donate her many other books and magazines. Her will was signed on September 13, 1917, witnessed by Dr. Julius Lackner, Alice Block, and her maid. As of September 1917, she seemed "normal" according to both Lackner and Block. Dr. Julius Lackner in response to how old she was noted "quite an old lady, about seventy."[31] A bill for doctor services itemizing frequent visits from November 13, 1917, to the day before she died, January 30, 1918, gives some indication of declining health. Most visits were "to 1 days services, 6 hours 50 cents per hour $3.00." The total bill of $54.00 was paid in full by the estate.[32]

Both Illinois Wesleyan University ($1,000) and McLean County Historical Society ($500) were left money in the last will and testament. Alvia K. Brown received money for legal services ($425) as well as a bequest ($2,000).

A forty-dollar granite marker was placed in Evergreen Cemetery. Paperwork indicates that it was ordered approximately one year after her death and paid for by Mr. Brown on December 26, 1918. Probate documents include a petition with regards to the books and magazines in the home. It indicates that shortly before her death Fitzwilliam had offered the books and magazines to the public library of Bloomington, but before the trustees could notify her of their interest, she had died. A friend agreed to gather the books and have them given to the public library of Bloomington, Illinois.

> Your petitioner further represents that the said decedent lived in the city of Bloomington for many years; that for more than eighteen (18) years she was a teacher in the public schools of that city and was, for several years, superintendent of the public schools of that city, and was very much attached to that city, not only on account of her long connection with its public schools, but because it had conferred upon her the unique distinction of appointing her as the first woman superintendent of public schools in the United States.[33]

Four brothers were listed as heirs at the time of death. Charles L. Raymond, Denver, Colorado; Frank C. Raymond, Gilbert, Arizona; Lyman H. Raymond, Kansas City, Missouri; and George W. Raymond Morris, Illinois. The letters of testamentary indicate Charles Raymond lived in Chicago at the home of Sarah Raymond. Margaret Kitson (probably the name of the maid) is also listed as residing at the home, 4824 Vincennes Avenue. This document notes the value of the whole estate at sixty-two thousand dollars.

This biography is the first extensive look at the nation's first female superintendent of city schools. Each piece of primary and secondary information offers important insight and understanding about the life and work of Sarah Raymond and implications for others in her day.

APPENDIX
Literature Review

> The range of what we think and do
> is limited by what we fail to notice
> And because we fail to notice
> that we fail to notice
> there is little we can do
> to change
> until we notice
> how failing to notice
> shapes our thoughts and deeds.
> —R. D. Laing, Scottish psychologist,
> poem in *Vital Lies, Simple Truths*

Linda Lawrence Hunt includes this poem as the chapter epigraph in the introduction to her book as it speaks to both the "neglect and the failure to recognize the importance of family stories of all people, not just the culturally privileged" and to academia for ignoring "the history of most women in the American story."[1] She focuses on the lost and silenced history of women at the end of the nineteenth century and the "negation through neglect" and silencing by intention. Hunt, writing about the lost story of one woman's journey with her daughter across America, enriched the record of the past by creating a "rag rug history." She pieces together the odyssey with "discards and remnants" from their own writings, letters, and newspaper accounts to create a fuller picture of American history. This creative work, much like the artistic rag rug products of earlier pioneer women, contributes to enriching the environment while preserving the past. The biography of Sarah Raymond serves as another wonderful example of a product expanding the record and understanding of American history and the everyday lives of ordinary women doing extraordinary things.

There are several ways one could look at similar projects in order to review the contributions to the field and establish a place for my study of Sarah Raymond Fitzwilliam. It is important to ground one's study in a larger scholarly context: thematically, methodologically, and contextually as explained later. One might consult published works on the topic, specifically, related topics, or topics thematically that are methodologically relevant. There have been no published books about the life and times of Sarah Raymond, so for a review of the literature one must consult related works. There have been some historical studies on women in education and educational leadership. There have been some case studies or biographies of noted and lesser-known women educators.

Teacher Narratives

University professors of education Preskill and Jacobvitz use teacher narratives from six genres as a way to discuss and explore what good teachers know and do. These stories draw from historical and contemporary experiences of teachers whose voices are known as well as those whose voices are less known. The stories are united under the following themes: social criticism, induction and apprenticeship, reflective practice, journey, hope, and freedom. The book contains excerpts from larger works of teacher stories thematically organized to illustrate ways these teachers have made a difference.[2] The story of Sarah Raymond will also serve to inspire future educators about challenging the status quo and making a difference. Preskill and Jacobvitz's use of voice and story to reveal truth and justice is compelling and inspiring.

Appendix

Greg Michie, a young teacher in his book *Holler if You Hear Me* (1999), tells the author's story together with the first-person stories of his students. Together they discover what it means to be a teacher and a student in urban America as they heighten awareness about teaching for social justice and making a difference for all.[3] The portrait of Sarah Raymond is framed by the voices and experiences of others around her. Just as Michie did in his book, I too have attempted to allow stories to tell the truth not only about education but also about society.

Case Study/Biography

Studies of women in education have universal lessons about the experience. Elizabeth Edwards wrote a historical account of three teacher training colleges in England during the early 1900s. Although the study of Sarah Raymond takes place in the United States, the themes that emerge in Edwards' book about teacher education are relevant as she investigates the lives and experiences of women. She adds to the literature a neglected aspect of women's experiences in the twentieth century. She opens her book with a quotation: "The challenge of the 'new' women's history, as it has been developed from the 1970s, has been to bring women 'back into' the historical record and to present them as individuals in their own right, as active agents in the making of history."[4] The individual stories of women and their contributions play a vital role in rewriting history and our feminist consciousness.

A scholar of educational leadership, Linda Lyman believes in making a difference and sees caring leadership as a key component. As a former school leader and current scholar of educational leadership she builds a three-year case study of a principal who together with parents, students, and staff built a circle of caring. The study of Ken Hinton is supported by theory and research on caring leadership as well.[5] The case study/biography of Sarah Raymond is grounded in larger theories of women in educational leadership in part to frame the picture.

Education professor Cordier, in her book *School Women of the Prairies and Plains* (1992), combines examples of historical teacher stories based on diaries and primary documents as case studies with a larger look at the educational and historical setting of school women.[6] This book offers interesting case study/biography modeling with broader contextual discussions much as I have framed my portrait of Sarah Raymond with the discussion of larger societal issues.

A historian and educator combined to write *Lives of Women Public Schoolteachers* (1995). Homes and Weiss look at how teachers taught and lived in the South, East, and on the West coasts, Midwest, and New England from the early 1800s to the 1950s. There is also a chapter on Normal Schools, which is of interest to this study as well.[7]

Educator Edwards tells the story of eight noted American women educators from 1820 to 1955 focusing on the educational practice of constructivism. In the eight women case studies, the author traces these practices in earlier times and makes note of the significant contributions they made not only to American education, but to society as a whole. Each chapter contains a biographical sketch of an educator's life and describes the reasons she embarked on her trailblazing endeavors, the hardships and hurdles encountered, her educational theories and methods, some criticisms voiced by contemporaries and present-day scholars, and her long-lasting achievements. It is the author's hope that readers, in learning about these educators' contributions and lives, will appreciate the inimitable courage they exhibited and the personal sacrifices they made in the face of official and societal discrimination and economic hardships. Reading the biographies of these remarkable women give, one insight into the times in which they lived and great inspiration for how to live more fully in our own.[8] Sarah Raymond, too, challenged the way things were being done, prescribed new and progressive curriculum approaches, experienced discrimination, and has left a lasting mark on the schools in Bloomington, Illinois.

Women Who Dared: Known and Lesser Known

There are also works about women who dared in various areas in the field of education. Numerous authors in numerous ways record their stories. There are straight narrative biographies as in the case of *The Power and Passion of M. Carey Thomas* (1994) by Helen Horowitz, the story of one woman president of Bryn Mawr College,[9] or Joyce Antler's book on *Lucy Sprague Mitchell* (1987). Mitchell was the founder of what is today known as the Bank Street College.[10] There are also books of essays and selected documents by educators including *Mary McLeod Bethune* (1999) edited by McCluskey and Smith, which reveals much about the advocate and spokeswoman for the oppressed from 1875 to 1955.[11] Writers Foner and Pacheco address the stories of three less-known female educators in antebellum America who taught black children. Although they taught in different parts of the country, many of their experiences were similar.[12] Nidiffer, professor of higher education, recounts the collective biography of the first deans of women offering illumination of women's history and higher education.[13] Although each of these examples is from a different writer, they are all united under the idea that the women portrayed were pioneers going against societal norms as they educated and led. This genre of study offers important insight into the challenges women faced as they dared to enter into educational leadership in a male-dominated society. There are wonderful examples of noted women who dared, but it is equally important to highlight less-known individuals who dared as well. This body of literature offers important models for my study. Sarah Raymond was a woman ahead of her time; daring to be superintendent when few women were, daring to challenge the issue of equal pay for women and men, daring to integrate the schools for black and white students, and daring to develop a progressive curriculum for Bloomington schools before national movements in that direction.

Feminist Historical Accounts

The following books are examples of works focusing on women's agency. These are stories of female educators challenging the establishment and making a difference in their field and their times. Whereas initial feminist work focused on the oppression of women, the field has emerged and themes reveal more nuances. Feminist historians acknowledge women's agency, uncover the considerable contributions of women, and celebrate their genuine accomplishments. By profiling women as educational activists, editors Crocco, Munro, and Weiler are challenging historical interpretations that have cast women as passive in the face of educational change.[14] Crocco and Davis, professors of social studies and education, edited two books together that examine the lives and work of women who forged a distinctive tradition of social education.[15]

Van Hover, another social studies education professor, published an interesting article examining the contributions to social education of Deborah Partidge Wolfe, a previously overlooked female African American educator.[16] Educational historian Blount tells the story of women and school leadership from 1873 to 1995. She uses statistical analysis, a historical approach, and feminist theory to focus on women in leadership with particular emphasis on the superintendency.[17] Patricia Carter, a scholar of women's studies, writes an interesting history of American women teachers uniting feminist ideology with the evolution of teaching. Carter suggested

> Feminism, in its many guises, guided teachers in their reform efforts. Its ideologies sustained them even though few ever utilized the term feminist as a self-referent. Whether teachers and their organizations identified as feminist or not, their attempts to make meaning of their lives within the gendered institution of schooling were liberatory. Intellectually motivated and keenly interested in higher wages, improved working

conditions, and expanded personal options, they were anything but the altruistic, self-abnegating servants delineated by early school planners.[18]

Although the term *feminist* comes about later than the period of Sarah Raymond, she embodies the attributes and attitudes of a feminist reformer, challenging injustice and discrimination she observed as a woman.

All of these authors offer important contributions to my study as they in various ways discuss the history of education from a feminist perspective. They are thus employing similar theoretical models to similar topics of study. My historical biography of Sarah Raymond is grounded in feminist traditions.

There are several models I draw from as I write the biography of the life, work, and legacy of Sarah Raymond Fitzwilliam. I tell her story using her writings and appropriate secondary sources. I frame her experiences within the larger context of the period and the role of women in education and leadership. I support this portrait of Sarah Raymond with the conceptual framework and appropriate method of inquiry. This literature review helps to put my study into a scholarly context and demonstrate what we already know and where we need to be heading.

Methodology and an Overview of Sarah Raymond's Life

The study of Sarah Raymond is important to the field of educational leadership and offers unique insight and perspectives to an underrepresented area of study. This study, much like *Pedagogies of Resistance* (1999), attempts to redress the marginalization of women in the history of education by offering important stories that among other things illuminate the gendered nature of educational change.[19] Belenky's work has helped to give voice to the field of women in educational research and build on the idea of women's ways of knowing by incorporating personal knowledge with knowledge gained from others.[20] The content and method of this study on Sarah Raymond sheds light on major societal and education related issues: history of education at the secondary and university level, woman's suffrage, and early women in leadership in and outside the classroom. The framework of this biography on Sarah Raymond is grounded in the ideas of critical theory, feminist theory, and the qualitative case study approach. Together with historical models and examples this study makes important contributions to the fields of social history and the history of education. It is a historical narrative biography based on primary and secondary research. The limitations to the study center on the availability of sources. Historical research is limited by sources and although there is sufficient evidence to develop a biography of Sarah Raymond there is not extensive documentation from her private letters. She was a very public person and active in the community. The arguments made are based on the evidence found.

Historiography

As in many fields, there has been an explosion of new ideas and a diversity of historical interpretations in recent years. Historiography, especially since 1945, has witnessed a progressive revolution. Particularly since the 1960s and 1970s, the field of social history has changed the profession and become an accepted area of research. This is not to say that the writing of everyday life did not exist prior to the 1960s, but social history has only more recently become accepted and respected. It is much more scientific and less anecdotal as well. The growth in social history also reflects a changing group of historians and a societal interest in hearing the stories of under-represented peoples and social reform, according to authors of social sciences education, Dynneson, Gross, and Berson.[21] In the history of education, the work by Theodore Sizer draws on the social history tradition as he looks at schooling in America.[22] My study about

Sarah Raymond draws from the areas of social history and women's history as it seeks to uncover a silent voice of the past.

Two historians, Furay and Salevouris, note a sobering fact, that although women constitute more than half of the human race, it was only after World War II that historians began to pay systematic attention to the role of women in history. For decades the male-dominated history profession systematically ignored them.[23] Along with the rise of women's history have emerged the new fields of gender studies, the history of gender relationships, histories of children and families, and gay and lesbian history. A women's historian, Scott points out the challenges within this important field. While attempting to recover an overlooked past and advance the cause of women's equality, what women's historians were discovering was that there is no such thing as a singular women's history relevant to all women everywhere.[24] The rich diversity of stories that characterized history writing in general at the turn of the century characterizes women's history today.

New fields emerge as the distinct discipline lines blur. Gilligan's study of moral judgment among women is considered to be the groundbreaking feminist work in social science and education research.[25] Noddings, a philosopher of education, expanded Gilligan's work with her moral theory of an ethic of caring. She notes: "If women's culture was taken more seriously in educational planning, social studies and history might have a very different emphasis. Instead of moving from war to war, ruler to ruler, one political campaign to the next, we would give far more attention to social issues."[26] She suggests more attention needs to be paid in the social studies curriculum to issues and practices that reflect caring (e.g., intergenerational responsibility and nonviolent conflict resolution).

In 1981, Congress declared the month of March to be women's history month, in part to help overcome the neglect of women in the history of our nation. As a special effort to redress the relative absence of women and their contribution and perspectives in written history, the field of women's history has developed and a month set aside. Martorella, a social science educator, discusses the congressional statement declaring March Women's History Month:

> American women of every race, class and ethnic background helped found the Nation in countless recorded and unrecorded ways as servants, slaves, nurses, nuns, homemakers, industrial workers, teachers, reformers, soldiers, and pioneers; and...served as early leaders in the forefront of every major progressive social change movement, not only to secure their own right of suffrage and equal opportunity but also in the abolitionist movement, the emancipation movement, the industrial labor union movement, and the modern civil rights movement; and...despite these contributions, the role of American women in history has been consistently overlooked and undervalued in the body of American history...[27]

Popular history, as it has come to be called, is another emerging approach to historiography in the information age or history targeted for mass audiences. Ultimately, it is important for historians to communicate their findings to a larger audience; otherwise they are serving no useful function in a society. On the other hand popular history can be a dangerous thing as good history is sacrificed for good entertainment. In fact, Furay and Salevouris mention the modern paradox that even as many critics lament America's increasing historical illiteracy, history has never clamored so insistently for our attention.[28] Over the last fifty years there has been an increase in the numbers of cheap paperback books, television programs, films, cable networks, and Internet sites. With the proliferation of the Internet and society's desire for the "inside" story, there is an increasing amount of poor history being distributed. The proliferation of "teacher" movies nicely illustrates this model. *Mr. Holland's Opus*, *The Dead Poets Society*, and *Dangerous Minds* are just a few examples of popular films about education.

In general the profession has expanded its traditional approach for conducting, interpreting, and writing history. The field has become more eclectic and diverse. It is important that previously underrepresented stories are emerging and that more people can personally connect with the past, but it is challenging at the same time, if universal stories of history are becoming lost. Historian Gilderhus recently noted that history no longer sets forth common stories that presumably speak for the identity and experience of all readers and that we no longer possess a past commonly agreed upon.[29]

The social sciences and literary criticism have sparked several other frameworks for history: postmodernism, deconstruction, semiotics, and structuralism/poststructuralism. Windschuttle, a modern historian, argues that the newly dominant theorists within the humanities and social sciences assert that it is impossible to tell the truth about the past or to use history to produce knowledge in any objective sense at all. They claim we can only see the past through the perspective of our own culture and hence what we see in history are our own interests and concerns reflected back at us. The central point upon which history was founded no longer holds: there is no fundamental distinction any more between history and myth.[30] This is a strong statement and a tension for the field and this project. Not all new historians argue for extreme forms of relativism. Appleby, Hunt, and Jacob contend that truths about the past are possible even if they are not absolute.[31] The postmodern tradition in history is similar to that which already exists in education research. One needs to only take a look at *Critical Pedagogy, the State, and Cultural Struggle* (1989) by Giroux and McLaren for a strong example of this theory in education.[32]

Methodology and Data Sources

Information was collected from several local archives and libraries to give voice to Sarah Raymond's story of leadership in the late 1800s. It is important to use case study and stories as a way to rediscover the past and forge a new future. While creating a portrait of Sarah Raymond, I frame it within important historical contexts supported by appropriate methodological inquiry and secondary sources. The conceptual framework I use helps to make this historical piece relevant to educators today by connecting to larger issues of teaching, learning, and leading—both past and present.

The archive at the Bloomington School District 87 central office contains many primary documents of the period and of the life work of Sarah Raymond. They have, for example, school board minutes, school records, school publications, and school scrapbooks. Many of these unpublished documents are the writings of Sarah Raymond. The McLean County Museum of History library houses several publications by Sarah Raymond, as well as archival material about her, the community, and her time period. The Illinois State University library and archive are also a source of helpful material. Since Sarah Raymond was an early graduate of the Illinois State Normal University many of the school's early publications offer insight into her early life and training as an educator. The local newspapers of Bloomington-Normal, Illinois, *The Pantagraph, The Bulletin,* and *The Leader,* were consulted as well since they offer important perspectives on Sarah Raymond's story and are located on microfilm at the Bloomington Public Library and Milner Library at Illinois State University. To continue building knowledge to support the study and give structure to the conceptual framework I use sources about the history of the Bloomington-Normal community, the period, schooling and teaching, women in educational leadership, and curriculum development.

Women and Education

Sarah Raymond was an excellent example of the "New Woman" who integrated Victorian virtues with an activist social role. The New Woman had an enhanced sense of self, gender, and mission. Although the "New Woman" reached her stride in the Progressive Era, she had antecedents in the 1870s and 1880s. Attributes often included middle-class status, educated, employed outside the home, not married, and active in single sex social organizations and associations. The single middle-class woman of the late nineteenth century had improved options for higher education, for professional employment, and for establishing supportive relationships with women outside the family.[33] Sarah Raymond remained single until she was fifty years old, left the superintendency, and moved to Boston. She was very close to Georgina Trotter and an active leader with many local and national organizations. Sarah Raymond was an excellent case study/biography of a woman in educational leadership ahead of her time.

Historical examples serve as powerful illustrations and, coupled with present day research in education, reveal much about the continuity and change in what is being taught, how it is being taught, and how it should be taught in American schools. Curricular decisions ebb and flow with the current political and social climate. It is important to understand this action/reaction process as it is relevant to more than just physics. This relationship is particularly visible in the discipline of history/social science of both the past and the present. The recent works by Carter and Doyle,[34] and Preskill and Jacobvitz on the importance of stories in the preparation of educators are very important to this field. The narratives of educators have greatly been over looked. The authors argue a shift in importance with the emergence of new lines of thinking.[35] Narratives, they argue, contribute beyond new lines of thinking and they discuss the following emerging new areas: (1) critics of politics of teaching, (2) the prominence of feminist thinking in teacher education, and (3) a growing appreciation of narrative as both a form of inquiry and a form of theory about teaching. Sarah Raymond serves as a wonderful case study biography of early pioneering efforts in curriculum, instruction, and administration as she was a leader in her field. Sarah Raymond wrote the *Rules and Regulations, Manual of Instruction to Teachers and Graded Course of Study of the Public Schools of Bloomington*. This was an attempt to formalize and perhaps professionalize the process of teaching, learning, and leading.

Women and Educational Leadership

Studies of women's leadership is a new outgrowth of more traditional studies of leadership and offers important perspectives, images, and values that have been missing from mainstream leadership theory and practice. Discussions of women in leadership are for the most part both absent and unique. Jackie Blount gives a historical account of women and school leadership in America from the common school era to the present. She illustrates how teaching emerges as women's work and school administration (superintendency) as men's work. This unique combination study of qualitative and quantitative sources explores how power in school employment has been structured unequally by gender.[36]

Spring illustrates similar points in his historical overview of *The American School: 1642–1996* (1997). The correlation between the employment of women and the pace of bureaucratization explains the relationship between the values inherent in bureaucratic organizations and attributes ascribed by nineteenth-century society to males and females. Men managed and women taught. This reflected nineteenth-century social patterns and was the basis of the hierarchical educational systems. Thus the function of women in the common school system was to be moral, nurturing, and loving teachers, guided and managed by men holding positions of authority as superintendents and principals.[37] This curriculum, school philosophy, and administrative structure are the foundation of our system today and a hard structure for women to challenge. The "one best system," which evolved in the common school era as a solution to the challenges of industrialization, urbanization, and immigration, was clearly not the best system for women leaving a lasting legacy of inequality of pay, status/power, and professionalism.

As the field of educational administration becomes more diversified, researchers are increasingly interested in studying these marginalized groups. And as new theories of education emerge, the topic of feminist images of leadership becomes a valid area of study. Larson and Murtadha have identified several important themes in the literature of women's leadership: (i) Male voice has been privileged and embedded within theories of knowledge and research methodologies long accepted as universal and neutral; (ii) issues of gender and access to administrative roles and gender equity in administration; and (iii) alternative images of effective leadership.[38]

Women construct and enact leadership in ways that depart from their male colleagues.[39] Grogan also found that women leaders often enact an ethic of care rooted in concerns for relationships rather than roles. Critical theorists look to improve society and challenge injustice. Feminist leadership research validates a different way of knowing and responding to moral dilemmas and civic responsibilities. Noddings argues that an ethic of care is fundamental for reframing and reorganizing schools.[40] This is a new relationship image of leadership and a departure from a hierarchical and a role-based image. Women's ways of leadership seem to embrace the new trends in leadership that were covered earlier: collective/democratic leadership, social justice/moral leadership, and community/caring leadership. In fact, Starratt discusses the concept of administering community. This is the very nature of combining care, justice, critique, and democracy.[41]

Only when the traditional hierarchical leadership structures of schools erode can new and more positive forms of leadership emerge. Only out of these new perspectives of what leadership is and the potential leadership has can women's leadership evolve and flourish. Caring, democratic, community, and collective leadership styles cannot be practiced within existing hierarchical, bureaucratic structures that reinforce the traditional, competitive, manipulative approaches to leadership. To accept the ideas included in feminist scholarship, traditional leaders will have to question much of what they have been taught so far.[42]

Lyman, Ashby, and Tripses identify several contemporary issues in the study of women's leadership: resolving cultural tensions; essentializing; including views of diverse women; feminist concerns; and new questions and emerging themes.[43] The cultural tensions that exist for women in leadership vary: for example, the female leader in a male sphere as Blount discusses.[44] Although legally there are no longer gender barriers, Smulyan recognizes the balancing act that female leaders play as a result of being a woman in a male role of power, authority, and leadership.[45] Essentializing and including views of diverse women are logical extensions of each other. The use of case study and individual stories is important to represent the underrepresented and highlight the various female leader values, visions, and experiences. Feminist concerns focus on language, interpretation, framework, role in society, research methodology, and type of leadership in general.

It is important to recognize the evolution of leadership and the subsequent emergence of women's leadership. There are many points of contrast and comparison between the traditional view of leadership, the new directions of leadership, and the current field of women's leadership. These leadership paradigms need to be understood as we move into the future of educational leadership.

Future of Educational Leadership and Women's Leadership

The future direction of education seems to be driven by external forces of standards and increased accountability. The question will be of how this imposed leadership model aligns with women's leadership. We see a complementary relationship between the new trends in leadership generally of democratic schools, a culture of caring, and learning communities and feminist leadership. One must look for these progressive elements in the imposed standards. There is evidence that collective leadership, community leadership, and teacher leadership can emerge out of the standards movement, but it will take effort on the part of practicing school leaders to not lose the progress that has been gained and revert to imposing control. Perhaps the new standards can serve as an ethics code for educators thus solidifying the role of ethical issues, caring, justice, and fairness in the role of school and leadership.

The value of life stories in leadership and in particular women's leadership has been discussed. These historical and contemporary biographies are imperative to understanding and reshaping school leadership. Stories reveal how educational leaders have struggled with issues related to the education of children and through this process have gained a sense of who they are and what they believe personally and professionally. Recognizing and giving voice is central to feminist ways of knowing with roots in the work of Gilligan, Noddings, Belenky, and many others.[46] Drawing from the historical work of Blount, it must be noted that women must shape the future structure of school and administrative leadership, if there is to be professional evolution of women's leadership. As Blount said, "If we continue to support schools that systematically distribute power unequally by sex and gender, we send a forceful message to students about women's worth, their potential, and their place in society."[47] Imposed standards and increased accountability is a reality, but how we interpret it and allow it to govern schools is open. If women are to maintain leadership or increase leadership roles in the future, they must not only join the profession but challenge its future, how it is written about, and the direction of administrative education programs. Madeleine Grumet in *Bitter Milk* (1988) challenges the reader and teacher to become empowered. "Stigmatized as 'women's work,' teaching rests waiting for us to reclaim it and transform it into the work of women."[48] The biography of Sarah Raymond Fitzwilliam sheds light on one woman's experience in a top position of educational leadership for the first time in American society. We must learn from past women who dared to lead, ahead of their times, to reclaim the profession, and ensure equality and justice for all.

NOTES

One   Introduction

1. Carolyn De Swarte Gifford, *Writing Out My Heart: Selections from the Journal of Frances E. Willard 1855–96* (Urbana: University of Illinois Press, 1995), xi.
2. Carol Ascher, Louise DeSalvo, and Sara Ruddick (eds), *Between Women: Biographers, Novelists, Critics, Teachers and Artists Write about Their Work on Women* (Boston: Beacon Press, 1984), xxii.
3. Sara Alpern, Joyce Antler, Elisabeth Israels Perry, and Ingrid Winther Scobie (eds), *The Challenge of Feminist Biography: Writing the Lives of Modern American Women* (Urbana: University of Illinois Press, 1992), 21.
4. Kathleen Weiler, *Country Schoolwomen: Teaching in Rural California 1850–1950* (Stanford: Stanford University Press, 1998), 1.
5. "Evergreen Cemetery (Bloomington, Illinois)," from Wikipedia, the free encyclopedia, accessed June 1, 2007; available from http://en.wikipedia.org/wiki/Evergreen_Cemetery_(Bloomington,_IL). This website offers information about the cemetery and noted members of the community buried there. There is, however, no mention of Sarah E. Raymond.
6. Charlie Schlenker, *Voices from the Past: Discovering Evergreen Cemetery* (Bloomington, IL: McLean County Historical Society, 2000), 29–30.
7. Ibid., 25–26.
8. *Illinois McLean County Cemeteries*, v:19 (Normal, IL: McLean County Genealogical Society, 2001), 143, 376, 472.
9. *The Chicago Evening Post*, February 1, 1918, 4.
10. *The Chicago Daily Tribune*, February 1, 1918.
11. Email from Marie Kroeger, Chicago, Illinois, to Monica Cousins Noraian, Bloomington, Illinois, July 21, 2005, July 28, 2005, and August 4, 2005.
12. Charles Capen, "Mrs. Sarah E. Raymond Fitzwilliam," *Journal of the Illinois State Historical Society* 11, no. 1 (April 1918): 81–82.
13. Sarah E. Raymond, Bloomington, Illinois, to Lyman Raymond, Seneca, Illinois, November 14? LS, Sarah Raymond Fitzgerald File, McLean County Historical Society Archives, Bloomington, Illinois. Letter donated by

Fitzgerald family member Arthur Thompson. (Exact year of letter unknown, but must have been prior to 1877 because it mentions Sarah Raymond's mother having a "very bad cold" at the time the letter war written. Sarah Raymond's mother died in 1877.)
14. Sarah E. Raymond and Jonathan Raymond, Bloomingotn, Illinois, to Lyman Raymond, Seneca, Illinois, December 11?, LS, Sarah Raymond Fitzgerald File McLean Country Historical Society Archives.
15. *The Alumni Quarterly of the I.S.N.U.* 7 no. 1 (February 1918): 3–6, 22.
16. Ibid.
17. *The Bloomington* (Illinois) *Pantagraph*, February 1, 1918, 10.
18. *The Bloomington* (Illinois) *Daily Bulletin*, February 1, 1918, 6.
19. Research was done at the McLean County Museum of History archives and conversations were had with Greg Koos, its director.
20. "FlickFact," *The Bloomington* (Illinois) *Pantagraph*, August 23, 2005, 2.
21. Barbara Finkelstein, "Revealing Human Agency: The Uses of Biography in the Study of Educational History," in *Writing Educational Biography: Explorations in Qualitative Research*, ed. Craig Kridel (New York: Garland Publishing, 1998), 46.
22. Mari Jo and Paul Buhle, *The Concise History of Woman Suffrage* (Urbana: University of Illinois Press, 1978), 1–45. They give an overview of the suffrage movement and its connection to other women's rights movements.
23. Ibid., 8.
24. Jo and Buhle, *The Concise History of Woman Suffrage*, 30.
25. Ibid., 35.

Two   The Early Years

1. *History of Kendall County* (Chicago: Munsell Publishing Company, 1914), 980.
2. Sarah Raymond contributed articles and biographical information about her family to the *History of Kendall County* published in 1914. Jonathan and Catherine Raymond biography sections, pp. 1037–1039, were written by Sarah E. Raymond Fitzwilliam.
3. *Transactions of the McLean County Historical Society, Bloomington, Illinois Volume II* (Bloomington: Pantagraph Printing and Stationery Company, 1903), 680.
4. *History of Kendall County*, 1037.
5. Ibid., 1037–1038.
6. Art Thompson, Manchester, Missouri, to Monica Cousins Noraian, Bloomington, Illinois, TLS, June 23, 2006. Written correspondence about Raymond family history.
7. *History of Kendall County*, 1039.
8. Ibid.

Notes

9. Notes from an unpublished family history from 1886 given to my by Art Thompson (his great-great grandparents were her parents). He and I have written letters and exchanged information about the family. He lives in Missouri.
10. *History of Kendall County*, 1038, and unpublished *Raymond Family Genealogy* from Art Thompson in the Raymond Files at the McLean County Historical Society, Bloomington, Illinois.
11. *History of Kendall County*, 1038.
12. The Kendall County Bicentennial Commission, *A Bicentennial History of Kendall County, Illinois* (Yorkville, IL: Kendall County Historical Society, 1976), 131.
13. Ibid., 134.
14. *History of Kendall County*, 766.
15. Ibid., 768.
16. Ibid.
17. Ibid., 767.
18. *A Bicentennial History of Kendall County, Illinois*, 142.
19. Ibid.
20. Edmund Hicks, "Chapter X," in *Hick's History of Kendall County* (Yorkville, IL: Record Print, 1927).
21. Art Thompson, Manchester, Missouri, to Monica Cousins Noraian, Bloomington, Illinois, TLS, June 23, 2006. Written correspondence about Raymond family history.
22. *History of Kendall County*, 980.
23. *The Bloomington* (Illinois) *Pantagraph*, February 1, 1918, 10.
24. *The Bloomington* (Illinois) *Daily Bulletin*, February 1, 1918, 6.
25. "Kendall County Country & Village Schools," accessed June 21, 2005; available at http://www.rootsweb.com/~ilkendal/.
26. Polly Welts Kaufman, *Women Teachers on the Frontier* (New Haven: Yale University Press, 1984), 39.
27. Ibid., 43.
28. *The Bloomington* (Illinois) *Daily Bulletin*, February 1, 1918, 6.
29. Charles Capen, "Sarah Raymond Fitzwilliam," *Journal of the Illinois State Historical Society* 11, no. 1 (April 1918): 81.
30. Census Data available at the Kendall County Historical Society.
31. Elmer Dickson, compiler and editor, *Teachers of Kendall County* (Chico, CA: Elmer Dickson, 2001), 76.
32. *Journal of the Illinois State Historical Society*, 11, no 1 (April, 1918).
33. *The Bloomington* (Illinois) *Pantagraph*, September 10, 1877, 4.
34. *The Bloomington* (Illinois) *Pantagraph*, November 15, 1877, 4.
35. *The Bloomington* (Illinois) *Pantagraph*, July 11, 1884, 4.
36. *The Bloomington* (Illinois) *Pantagraph*, February 1, 1918, 10.
37. Erma Scarlette, "A Historical Study of Women in Public School Administration from 1900–1977." EdD dissertation, The University of North Carolina at Greensboro, 1979, 4–6.

38. June Edwards, *Women in American Education, 1820–1955: The Female Force and Educational Reform* (Westport, CT: Greenwood Press, 2002), xi–xvii.

Three The Illinois State Normal University Years

1. Homer Hurst, *Illinois State Normal University and the Public Normal School Movement* (Nashville: George Peabody College for Teachers, 1948), 22.
2. Jacob Hasbrouck, *History of McLean County* (Topeka: Historical Publishing Company, 1924), 193.
3. Christine Ogren, *The American State Normal School: An Instrument of Great Good* (New York: Palgrave, 2005), 57.
4. Sandra Harmon, "The Voice, Pen and Influence of Our Women are Abroad in the Land: Women and the Illinois State University, 1857–1899," in *Nineteenth-Century Women Learn to Write*, ed. Catherine Hobbs (Charlottesville: University Press of Virginia, 1995), 94–95.
5. Burt Loomis, *The Educational Influence of Richard Edwards* (Nashville: George Peabody College for Teachers, 1932), 81–82.
6. *Catalogue of the State Normal University, for the academic year ending 1862,* Illinois State University Archives, Normal, Illinois.
7. Consulted catalogues for all the years Sarah Raymond attended ISNU, 1862–1866, in the Illinois State University Archives. *Catalogues of the State Normal University, for the academic years ending 1862, 1863, 1864, 1865, 1866.*
8. *Catalogue of the State Normal University, for the academic year ending 1862.*
9. Harmon, "The Voice, Pen and Influence of Our Women," 94–95.
10. Ibid., 95; Jurgen Herbst, *And Sadly Teach: Teacher Education and Professionalization in American Culture* (Madison: University of Wisconsin Press, 1989), 11; "Address of Edwin C. Hewett, LL.D.," in *A History of the Illinois State Normal University, Normal, Illinois,* ed. John Cook and James McHugh (Bloomington, IL: Pantagraph Printing, 1882), 209.
11. Hurst, *Illinois State Normal University and the Public Normal School Movement,* 24.
12. Ibid., 25.
13. Loomis, *The Educational Influence of Richard Edwards,* 85.
14. Ibid.
15. Ibid., 85–86.
16. John Freed, *Educating Illinois: Illinois Sate University, 1857–2007* (Virginia Beach, VA: The Donning Company Publishers, 2009), 101.
17. *Catalogue of the State Normal University, for the Academic Year Ending 1862.*
18. Illinois State Normal University, "*Alumni Register 1860–1927*," *The Normal School Quarterly* series 26, no. 104 (July, 1927): 271–272.
19. Hurst, *Illinois State Normal University and the Public Normal School Movement,* 25.
20. Ibid., 26.

21. Charles Harper, *Development of the Teachers College in the United States with Special Reference to the Illinois State Normal University* (Bloomington: McKnight & McKnight, 1935), 68.
22. Hurst, *Illinois State Normal University and the Public Normal School Movement,* 25–26.
23. Loomis, *The Educational Influence of Richard Edwards,* 85.
24. John Freed, *Educating Illinois: Illinois State University, 1857–2007* (Virginia Beach: The Donning Company Publishers, 2009), 118.
25. Illinois State Normal University, *Semi-Centennial History of the Illinois State Normal University, 1857–1907.* Prepared under a committee of the faculty (Normal, Illinois, 1907), 78.
26. Loomis, *The Educational Influence of Richard Edwards*, 89–90.
27. Ogren, *The American State Normal School,* 31.
28. Loomis, *The Educational Influence of Richard Edwards*, 90–91.
29. Illinois State Normal University, *Semi-Centennial History of the Illinois State Normal University,* 73.
30. Loomis, *The Educational Influence of Richard Edwards*, 91–92.
31. Ibid., 92–93.
32. *Wrightonia Record Book Minutes*, unpublished document, Illinois State University Archives.
33. Freed, *Educating Illinois*, 135.
34. *Wrightonia Record Book Minutes*, unpublished document, Illinois State University Archives.
35. Freed, *Educating Illinois*, 136.
36. *Wrightonia Record Book Minutes*, unpublished document, Illinois State University Archives.
37. Helen Rudd, unpublished diary, January 1, 1865–January 7, 1872, Illinois State University Archives.
38. Helen Rudd, unpublished diary, Wednesday January 4, 1865, Illinois State University Archives.
39. Helen Rudd, unpublished diary, May 17, 1866, Illinois State University Archives.
40. Helen Rudd, unpublished diary, May 19, 1866, Illinois State University Archives.
41. *Wrightonia Record Book Minutes.*
42. Loomis, *The Educational Influence of Richard Edwards*, 82.
43. Ibid.
44. "The State Normal," *The Bloomington* (Illinois) *Pantagraph*, February 1, 1875, 3.
45. Graduation program, *Order of Exercises at the seventh commencement of the Illinois State Normal University, Thursday, June 28th 1866*, Pantagraph Print, Illinois State University Archives.
46. Richard Edwards, ISNU Class of 1866 Graduation Address, June 27, 1866, 1–33, ISU Archives, unpublished Edwards papers.

47. "The Illinois State Teachers Association," *Chicago Tribune*, December 31, 1863, 3.
48. Ibid.
49. "Illinois State Teachers' Institute-Third Day," *The Bloomington* (Illinois) *Daily Pantagraph*, August 10, 1871, 4.
50. Illinois State Normal University, *Semi-Centennial History of the Illinois State Normal University*, 122–123.
51. "Educational," *The Bloomington* (Illinois) *Pantagraph*, June 11, 1878, 3.
52. "Teachers' Institute," *The Bloomington* (Illinois) *Pantagraph*, July 10, 1874, 4.
53. "They Who Teach: the County Institute Now in Session at the High School," *The Bloomington* (Illinois) *Pantagraph*, August 10, 1878, 3.
54. "Illinois State Teachers' Institute- Second Day," *The Bloomington* (Illinois) *Daily Pantagraph*, August 9, 1871, 4.
55. "Illinois State Teachers' Institute – Fifth Day," *The Bloomington* (Illinois) *Daily Pantagraph*, August 12, 1871, 4.
56. *Catalogue of the Illinois State Teachers Institute*, held at the Normal University August 1869 (Normal, Illinois: Pantagraph Book and Job Office, 1869), Illinois State University Archives.
57. *Catalogue of the Illinois State Teachers Institute*, held at the Normal University August 1871 (Normal, Illinois: Leader Company Print, 1871), Illinois State University Archives.
58. *Bloomington* (Illinois) *Daily Pantagraph*, June 24, 1897, 8.
59. John Cook and James McHugh, *A History of the Illinois State Normal University* (Normal, Illinois: Pantagraph Printing, 1882), 227–235.
60. Helen Marshall, *Grandest of Enterprises* (Normal, IL: Illinois State University, 1956), 207.
61. *The Illinois State Normal University Vidette*, June 1897, 21.
62. *The Illinois State Normal University Index, 1897*, 145.
63. "Anniversary Exercises at Normal," *Chicago Daily Tribune,* June 24, 1897, 8.
64. Illinois State Normal University, *Semi-Centennial History of the Illinois State Normal University*, 1857–1907 (Normal, IL: Illinois State University, 1907), 188–189.
65. Sarah E. Raymond Fitzwilliam, "The Old Plank Walk," in *Semi-Centennial History of the Illinois Normal University*, 223–225.
66. Ibid.
67. Ibid.
68. Ibid., 224.
69. Freed, *Educating Illinois*, 77.
70. *The Alumni Quarterly* (February 1914) and *The Illinois State Normal University Vidette* (February 11 and 18, 1914).
71. Sarah E. Raymond, Chicago, Illinois, to President David Felmley, Normal, Illinois, LS, January 30, 1914, Illinois State University Archives, Metcalf Dedication Files.
72. Ibid.
73. *The Alumni Quarterly* (May 1916), 28, and *Normal School Quarterly* (1927), 8.

74. Sandra Harmon, "The Voice, Pen and Influence of Our Women are Abroad in the Land: Women and the Illinois State University, 1857–1899," in *Nineteenth-Century Women Learn to Write*, ed. Catherine Hobbs (Charlottesville: University Press of Virginia, 1995), 99.
75. "Sarah E. Raymond Fitzwilliam," *The Alumni Quarterly of the I.S.N.U.* 7, no. 1 (February 1918): 22.
76. Ibid.
77. Nancy Hoffman, *Woman's "True" Profession: Voices from the History of Teaching* (Cambridge: Harvard Education Press, 2003); Christine Ogren, The *American State Normal School: An Instrument of Great Good* (New York: Palgrave, 2005); and Sari Knopp Biklen, *School Work: Gender and the Cultural Construction of Teaching* (New York: Teachers College Press, 1995).

Four   Teacher and Principal of Bloomington Schools

1. *Transactions of the McLean County Historical Society: Bloomington, Illinois, Vol. II* (Bloomington, Illinois: Pantagraph Printing and Stationery Co., 1903), 598.
2. "Close of a Career: Faithful School Work of Miss Sarah E. Raymond," *The Bloomington* (Illinois) *Leader*, July 15, 1892.
3. "The Public Schools," *Bloomington* (Illinois) *Weekly Pantagraph*, September 15, 1870, 3.
4. Sarah Raymond Fitzwilliam, "History of the Public Schools of Bloomington," *Transactions of the McLean County Historical Society Volume II* (Bloomington: Pantagraph Printing, 1903), 61.
5. "A Speech of War," *Bloomington* (Illinois) *Daily Pantagraph*, January 12, 1871, 4.
6. "The War Still Wages," *The Bloomington* (Illinois) *Daily Pantagraph*, January 13, 1871, 4.
7. *The Bloomington* (Illinois) *Daily Pantagraph*, January 21, 1871, 4.
8. *The Bloomington* (Illinois) *Daily Pantagraph*, January 14, 1871, 4.
9. *The Bloomington* (Illinois) *Daily Pantagraph*, January 23, 1871, 4.
10. "Colored Children in the Public Schools," *The Bloomington* (Illinois) *Daily Pantagraph*, June 13, 1871, 4; and *The Bloomington* (Illinois) *Weekly Pantagraph*, June 16, 1871, 4.
11. Robert McCaul, *The Black Struggle for Public Schooling in Nineteenth-Century Illinois* (Carbondale: Southern Illinois University Press, 1987), 128.
12. "Cases Decided," *The Bloomington* (Illinois) *Daily Pantagraph*, June 23, 1871, 4.
13. "Meeting of the Board of Education," *The Bloomington* (Illinois) *Weekly Pantagraph*, November 6, 1874, 3.
14. Fitzwilliam, "History of the Public Schools of Bloomington," 61.
15. John Freed, *Educating Illinois: Illinois State University, 1857–2007* (Virginia Beach: The Donning Company Publishers, 2009), 114.

16. "Close of a Career: Faithful School Work of Miss Sarah E. Raymond," *The Bloomington* (Illinois) *Leader,* July 15, 1892.
17. Fitzwilliam, "History of the Public Schools of Bloomington," 62.
18. *The Bloomington* (Illinois) *Daily Pantagraph*, May 15, 1871, 4.
19. Bloomington Teachers Association, Record Book, February 24, 1872, unpublished, 35, Bloomington School District 87 Archives, Bloomington, Illinois.
20. Bloomington Teachers Association, Record Book, December 13, 1873, unpublished, 57–58, Bloomington School District 87 Archives, Bloomington, Illinois.
21. "Compulsory Education," *The Bloomington* (Illinois) *Daily Pantagraph*, August 26, 1871, 2.
22. "Let Pupils Take Their Books Home," *The Bloomington* (Illinois) *Daily Pantagraph,* May 11, 1871, 4.
23. "High School Commencement," *The Bloomington* (Illinois) *Daily Pantagraph,* June 15, 1871, 4, and June 17, 1871, 4.
24. "The Employment of Lady Teachers in the Bloomington Schools," *The Bloomington* (Illinois) *Daily Pantagraph,* May 25, 1871, 2.
25. "Jacob, Marsh, Trotter: A Practical Demonstration of Bloomington's Belief in Woman's Suffrage," *The Bloomington* (Illinois) *Weekly Pantagraph*, March 27, 1874, 3.
26. "A New Departure," *The Bloomington* (Illinois) *Pantagraph*, April 1, 1874, 4.
27. "Miss Trotter Accepts," *Bloomington* (Illinois) *Pantagraph*, April 2, 1874, 4.
28. "Who is Who," *Bloomington* (Illinois) *Pantagraph*, April 3, 1886, 4.
29. "The Board Election," *The Bloomington* (Illinois) *Pantagraph,* May 3, 1886, 3.
30. David Tyack and Elisabeth Hansot, *Learning Together: A History of Coeducation in American Schools* (New Haven: Yale University Press, 1990), 87.
31. Dortha Tompkins, *District Eighty Seven*, Bloomington, Illinois, 1976, unpublished manuscript, 76, Bloomington District 87 Archives, Normal, Illinois; McLean County Probate Will Records and Estate Inventory, Illinois Regional Archives Depository System (IRAD), ISU Archives, Normal, Illinois.
32. Jonathan Raymond, Will and Probate, IRAD ISU archives.
33. "The Trotter Name," *The Bloomington* (Illinois) *Pantagraph,* May 30, 1911, 4.
34. Trotter Family Files, McLean County Historical Society, Bloomington, Illinois.
35. Jeanne Weimann, *The Fair Women* (Chicago: Academy, 1981), 290.
36. Carl Schlenker, *Voices from the Past: Discovering Evergreen Cemetery* (Bloomington, IL: McLean County Historical Society, 2000), 26.
37. "A Welcome Release," *The Bloomington* (Illinois) *Daily Bulletin*, February 9, 1892, 1.
38. *Bloomington and Normal City Directory for 1891* (Bloomington: Pantagraph Printing, 1891), 394.
39. "Visited Many Odd Corners of Europe," *The Bloomington* (Illinois) *Pantagraph*, November 29, 1913.

Notes

40. Sarah Deutsch, *Women and the City: Gender, Space, and Power in Boston, 1870–1940* (New York: Oxford University Press, 2000), 104–105.
41. "On the Threshold: From the School Room into the Battle of Life," *The Bloomington* (Illinois) *Pantagraph*, June 13, 1874, 4.
42. Reception invitation found in the Jackman family memorabilia file at the McLean County Historical Society, Bloomington, Illinois.
43. *The Aegis* (Bloomington: Frank Miller Company Publishers, 1911); located at the McLean County Historical Society, Bloomington, Illinois.
44. *History of Kendall County* (Chicago: Munsell Publishing Company, 1914), 982–983.
45. Thomas Glass, *The History of Educational Administration Viewed Through its Textbooks* (Lanham, MD: Scarecrow Education, 2004), 3.
46. Ibid., 13.

Five   Superintendent of Bloomington Schools

1. Sarah Raymond Fitzwilliam, "History of the Public Schools of Bloomington," *Transactions of the McLean County Historical Society Volume II* (Bloomington: Pantagraph Printing, 1903), 62.
2. "A New School Superintendent," *The Bloomington* (Illinois) *Pantagraph*, July 2, 1874, 4.
3. Ibid.
4. Ibid.
5. "School Matters," *The Bloomington* (Illinois) *Pantagraph,* July 7, 1874, 4.
6. "School Superintendency," *The Bloomington* (Illinois) *Pantagraph*, July 9, 1874, 4.
7. *Transactions of the McLean County Historical Society: Bloomington, Illinois, Vol. II* (Bloomington, Illinois: Pantagraph Printing and Stationery, 1903), 90.
8. "Our City Schools: Energetic Work and Thorough Discipline, the Old Fashioned Pedagogue and the New Era of Public Education," *The Bloomington* (Illinois) *Pantagraph,* October 27, 1874, 4.
9. Bloomington Teachers Association, Record Book, February 24, 1872, 64, unpublished, Bloomington School District 87 Archives, Normal, Illinois.
10. "Monthly Meeting of Teachers," *The Bloomington* (Illinois) *Weekly Pantagraph*, November 6, 1874, 3.
11. "What They Do at Teacher's Meetings," *The Bloomington* (Illinois) *Pantagraph*, February 1, 1875, 4.
12. Bloomington Teachers Association, Record Book, 77.
13. Homer Hurst, *Illinois State Normal University and the Public Normal School Movement* (Nashville: George Peabody College for Teachers, 1948), 27.
14. "Why the Schoolrooms Were Cold," *The Bloomington* (Illinois) *Pantagraph*, January 26, 1875, 4.
15. *The Bloomington* (Illinois) *Pantagraph*, February 5, 1875, 4, and February 6, 1875, 3.

16. "Our Schools: Meeting of the School Board Yesterday Afternoon," *The Bloomington* (Illinois) *Pantagraph*, March 2, 1875, 4.
17. "Board of Education," *The Bloomington* (Illinois) *Pantagraph*, February 2, 1875, 4.
18. "Educational: a Lively Meeting of the Board of Education," *The Bloomington* (Illinois) *Pantagraph,* April 13, 1875, 4.
19. "Among The Schools," *The Bloomington* (Illinois) *Pantagraph*, 3 April 1875, p. 3.
20. "The End," *The Bloomington* (Illinois) *Pantagraph*, June 21, 1878, 4.
21. *The Bloomington* (Illinois) *Pantagraph*, June 10, 1875, 4.
22. "Ended: Another Year's Work by Pupil and Teacher Gone Upon the Record," *The Bloomington* (Illinois) *Pantagraph*, June 11, 1875, 4.
23. "Our Schools: Special Meeting of the Board of Education Last Evening, Miss Raymond Re-Elected to the Head of the System," *The Bloomington* (Illinois) *Pantagraph*, July 13, 1875, 4.
24. "Our Schools: Who Are to General the Army of Teachers Next Year," *The Bloomington* (Illinois) *Pantagraph*, June 15, 1875, 4.
25. "Bloomington School Affairs," *Chicago Daily Tribune*, July 13, 1875, 8.
26. Ibid.
27. "School Superintendent," *The Bloomington* (Illinois) *Pantagraph*, October 8, 1875, 4.
28. "Education: The Bloomington School Board in Their Regular Monthly Session," *The Bloomington* (Illinois) *Pantagraph*, October 5, 1875, 4; and "Schools: Those of Bloomington Are in the Best of Working Order," *The Bloomington* (Illinois) *Pantagraph*, December 7, 1875, 4.
29. "Education: The Bloomington School Board in Their Regular Monthly Session," 4.
30. Fitzwilliam, "History of the Public Schools of Bloomington," 64.
31. Sarah E. Raymond, *Fifth Annual Report of the Bloomington Public Schools for the Year Ending June 10th, 1881* (Bloomington, IL: Bulletin Printing and Publishing, 1881), 14.
32. "Board of Education," *The Bloomington* (Illinois) *Daily Bulletin*, October 5, 1881.
33. Sarah E. Raymond, *Sixth Annual Report of the Bloomington Public Schools for the Year Ending June 8th, 1882* (Bloomington, IL: Bulletin Printing and Publishing, 1882), 15.
34. Ibid.
35. Ibid., 17.
36. Ibid..
37. Ibid.
38. Ibid., 41–42.
39. Ibid., 5.
40. Sarah E. Raymond, *Seventh Annual Report of the Bloomington Public Schools for the Year Ending June 8th, 1883* (Bloomington, IL: Bulletin Co., Printers and Publishers, 1883), 5.

Notes

41. Ibid., 11.
42. Sandra Harmon, "The Voice, Pen and Influence of Our Women are Abroad in the Land: Women and the Illinois State University, 1857–1899," in *Nineteenth-Century Women Learn to Write*, ed. Catherine Hobbs (Charlottesville: University Press of Virginia, 1995), 87; and *Proceedings of the Board of Education*, December 8, 1897 (Springfield, IL: Phillips Brothers, 1898), 6–7.
43. Carl Degler, *At Odds: Women and the Family in America from the Revolution to the Present* (Oxford: Oxford University Press, 1980), 309.
44. Raymond, *Seventh Annual Report of the Bloomington Public Schools*, 16.
45. Dortha Tompkins, *District Eighty Seven, Bloomington, Illinois*, 1976, unpublished manuscript, 104.
46. Raymond, *Seventh Annual Report of the Bloomington Public Schools*, 16.
47. Ibid., 17.
48. Ibid.
49. Sarah E. Raymond, *Eighth, Ninth and Tenth Annual Reports of the Bloomington Public Schools for the Years 1884, 1885 and 1886* (Bloomington, IL: Leader Publishing Col, Printers, 1886), 10–11.
50. Sarah E. Raymond, *Fifteenth Annual Report of the Bloomington Public Schools for the Year 1890–1891* (Bloomington, IL: The Leader, Printers, 1891), 11.
51. "German to be Taught in the Public Schools," *The Bloomington* (Illinois) *Daily Pantagraph*, August 30, 1871, 4.
52. Raymond, *Eighth, Ninth and Tenth Annual Reports of the Bloomington Public Schools*, 31.
53. Ibid., 32.
54. Sarah E. Raymond, *Eleventh Annual Report of the Bloomington Public Schools for the Years 1886 and 87* (Bloomington, IL: Leader Publishing Co., Printers, 1887), 6.
55. Ibid., 9.
56. Ibid., 17.
57. Sarah E. Raymond, *Twelfth Annual Report of the Bloomington Public Schools for the Year 1887 and 1888* (Bloomington, IL: Leader Publishing Co., 1888), 3 and 6.
58. Sarah E. Raymond, *Thirteenth Annual Report of the Bloomington Public Schools for the Year 1888 and 1889* (Bloomington, IL: Pantagraph Printing and Stationery Co., 1890), 19.
59. Sarah E. Raymond, *Sixteenth Annual Report of the Bloomington Public Schools for the Year 1891–1892* (Bloomington, IL: The Leader, Printers, 1892), 34–35.
60. Record Book, Board of Education, June 1891, Bloomington District 87 Archives, Normal, Illinois, 68.
61. There were many front-page articles about the contest over the three-month period but here are a few of interest. *The Bloomington* (Illinois) *Daily Bulletin*, January 6, 1892, February 2, 1892, April 29, 1892, and May 3, 1892.
62. "The Race Decided: the Bulletin's Great Teachers' Contest Magnificently Finished," *The Bloomington* (Illinois) *Daily Bulletin*, 3 May 1892, p. 1.

63. "Our Schools," *The Bloomington* (Illinois) *Pantagraph*, September 10, 1877, 4.
64. "Bloomington Schools: the First Annual Report of the Superintendent," *The Bloomington* (Illinois) *Pantagraph*, 10 September 1877, p. 3.
65. Ibid.
66. *The Bloomington* (Illinois) *Pantagraph*, September 2, 1878, 3.
67. "School-days Ended: The Annual Commencement of the High School of Bloomington," *The Pantagraph*, June 14, 1878, 4.
68. Raymond, *Fifth Annual Report of the Bloomington Public Schools*, 13.
69. Ibid.
70. Ibid.
71. Ibid.
72. Raymond, *Sixteenth Annual Report of the Bloomington Public Schools*, 6–7.
73. Steven Tozer, *School and Society: Historical and Contemporary Perspectives* (Boston: McGraw Hill, 2006), 38.
74. Sarah Raymond, *Rules and Regulations, Manual of Instruction to Teachers, and Graded Course of Study of the Public Schools of Bloomington* (Bloomington, IL: Bulletin Printing, 1883), 60.
75. Ibid., 92.
76. James Loewen, *The Lies My Teacher Told Me: Everything Your American History Textbook Got Wrong* (New York: Simon & Schuster, 1996).
77. Raymond. *Rules and Regulations*, 97.
78. Ibid., 197.
79. Ibid., 226.
80. Thomas Dynneson, Richard Gross, and Michael Berson, *Designing Effective Instruction for Secondary Social Studies* (Upper Saddle River, NJ: Merrill Prentice Hall, 2003).
81. "Bloomington Schools: The First Annual Report of the Superintendent," *The Bloomington* (Illinois) *Pantagraph*, September 10, 1877, 3.

Six The Resignation

1. *Sixteenth Annual Report of the Bloomington Public Schools for the year 1891–1892*. Prepared by Superintendent of City Schools, Sarah E. Raymond, The Leader, Printers, Bloomington, IL, 1892, 21.
2. Ibid., 23.
3. Ibid., 24.
4. "Educational Gossip," *Chicago Daily Tribune*, September 24, 1892, 13.
5. "Superintendent of Bloomington Schools," *Chicago Daily Tribune*, July 19, 1892, 1.
6. "Prof. E.M. Van Petten Chosen," *The Bloomington* (Illinois) *Pantagraph*, July 19, 1892, 7.
7. "Law for the Ladies," *The Bloomington* (Illinois) *Leader*, March 18, 1892, 7.

8. "Will Take a Hand," *The Bloomington* (Illinois) *Weekly Bulletin*, April 1, 1892, 2.
9. "To Prepare for Voting," *The Bloomington* (Illinois) *Leader*, March 28, 1892, 5.
10. *The Bloomington* (Illinois) *Leader*, April 2, 1892, 7.
11. "Mr. Thomas Accepts," *The Bloomington* (Illinois) *Leader*, March 28, 1892, 5. Mr. Thomas accepts the nomination for reelection to the School Board after published letter of support signed by many including F. J. Fitzwilliam (future husband of Sarah Raymond and Charles Capen, future author of her obituary for the Illinois Historical Society). Mr. Thomas notes he had been a member of the board for the past thirteen years and "if elected will do everything in my power to advance the best interests of our schools."
12. "The School Election," *The Bloomington* (Illinois) *Sunday Bulletin*, April 3. 1892, 3.
13. "The Merry War Is On," *The Bloomington* (Illinois) *Daily Bulletin*, April 4, 1892, 1.
14. Carolyn DeSwarte Gifford and Amy Slagell, ed., *Let Something Good Be Said: Speeches and Writings of Frances E. Willard* (Urbana: University of Illinois Press, 2007), xxxii.
15. "Battle of Ballots," *The Bloomington* (Illinois) *Leader*, April 5, 1892, 8.
16. "The Merry War Is On," 1.
17. Wanda Hendricks, *Gender, Race, and Politics in the Midwest: Black Club Women in Illinois* (Bloomington, IN: Indiana University Press, 1998), 130–131.
18. Ibid., 83.
19. "Women Against Women," *The Bloomington* (Illinois) *Leader*, April 5, 1892, 4.
20. "Status of the Schools," *The Bloomington* (Illinois) *Leader*, 6 April 1892, p. 4.
21. *The Bloomington* (Illinois) *Leader*, April 5, 1892, 4; letter to the editor by James Shaw, *The Bloomington* (Illinois) *Daily Bulletin*, April 7, 1892, 2; "Yesterday's Election," *The Bloomington* (Illinois) *Daily Bulletin*, April 5, 1892, 4; letter to the editor titled "The Issue Today," *The Bloomington* (Illinois) *Pantagraph*, April 4, 1892.
22. Sarah Raymond Fitzwilliam, "History of the Public Schools of Bloomington from 1825–1892," *Transactions of the McLean County Historical Society: School Record of McLean County and Other Papers, Volume II* (Bloomington, IL: Pantagraph Printing and Stationery Co., 1903), 65.
23. Record Book—Board of Education November 3, 1890–June 22, 1899. May 2, 1892. 113–114. School Board Record Books are housed in the District 87 Archives in Bloomington, Illinois
24. Record Book—Board of Education November 3, 1890–June 22, 1899. June 6, 1892. 116–117.
25. Record Book—Board of Education November 3, 1890–June 22, 1899. July 5, 1892. 128–129.
26. Ibid.

27. Record Book—Board of Education November 3, 1890–June 22, 1899. July 15 and 18, 1892. 131.
28. "Close of a Career," *The Bloomington* (Illinois) *Leader*, July 15, 1892.
29. "Contests in Illinois: Women Figure Prominently in the City Election at Bloomington," *Chicago Daily Tribune*, April 4, 1893, 3.
30. Humphrey Desmond, *The A.P.A. Movement* (New York: Arno Press, 1969), 7–9.
31. Donald Kinzer, *An Episode in Anti-Catholicism: The American Protective Association* (Seattle: University of Washington Press, 1964), 195.
32. Ibid., 64.
33. Ibid., 81.
34. David Tyack and Elisabeth Hansot, *Managers of Virtue: Public School Leadership in America, 1892–1980* (New York: Basic Books, 1982), 72–83.
35. Fitzwilliam, "History of the Public Schools of Bloomington," 65.

Seven Leading beyond the Schools: Community Involvement in Bloomington, Boston, and Chicago

1. "Close of a Career," *Bloomington* (Illinois) *Leader*, July 15, 1892.
2. Steven Rockefeller, *John Dewey: Religious Faith and Democratic Humanism* (New York: Columbia University Press, 1991), 229.
3. Allen Davis, *American Heroine: The Life and Legend of Jane Addams* (New York: Oxford University Press, 1973), 97.
4. Sarah E. Raymond, "The Illinois School Mistresses' Club," *Chicago* (Illinois) *Daily Tribune*, November 3, 1888, 12.
5. Joan Smith, *Ella Flagg Young: Portrait of a Leader* (Ames: Educational Studies Press, 1976), 50.
6. John Freed, *Educating Illinois: Illinois State University, 1857–2007* (Virginia Beach: The Donning Company Publishers, 2009), 147.
7. "Eight Hundred Schoolmarms," *Chicago* (Illinois) *Daily Tribune*, November 24, 1888, 1.
8. Elizabeth Cady Stanton, *History of Woman Suffrage* (New York: Arno Press, 1969), 578 and 585.
9. *Collection of Memorial Tributes to Adlai E. Stevenson*, 76–77, Library of Congress, electronic collection.
10. Ivan Light, *This Blooming Town: A Sketch of Bloomington, Illinois* (Bloomington, IL: Light House Press, 1956), 40.
11. *History of Kendall County* (Chicago: Munsell Publishing Company, 1914), 981.
12. It is possibly an error in the original document that stated that she knew Jane Austin. More probable would be that she knew Jane Addams.
13. "Married in Boston," *The Bloomington* (Illinois) *Pantagraph*, June 25, 1896, 7.

Notes

14. Jane Addams, *Twenty Years at Hull-House with Autobiographical Notes* (Boston: Bedford/St. Martin's, 1999), 212–217.
15. Allen Davis, *American Heroine: The Life and Legend of Jane Addams* (New York: Oxford University Press, 1973), 128.
16. "Bloomington in Chicago," *The Bloomington* (Illinois) *Pantagraph*, February 4, 1897, 8.
17. "Dust to Dust," *The Bloomington* (Illinois) *Pantagraph*, April 26, 1893, 7.
18. "Obituary, Captain Frank J. Fitzwilliam," *Chicago* (Illinois) *Tribune*, December 25, 1899, 5.
19. Stacy Cordery, "Women in Industrializing America," in *The Gilded Age: Essays on the Origins of Modern America,* ed. Charles Calhoun (Wilmington: Scholarly Resources Inc., 1996), 111–135.
20. Anne Firor Scott, *Natural Allies: Women's Associations in American History* (Chicago: University of Chicago Press, 1991), 184–189.
21. "News of Chicago Clubs and the Society World," *Chicago* (Illinois) *Daily Tribune*, December 13, 1913, 17.
22. *Journal of the Illinois State Historical Society*, VII, no. 3 (October 1914): 296; and two different volumes of *Transactions of the Illinois State Historical Society for the year 1904 and for the year 1918*.
23. Chicago Historical Society, Hyde Park Travel Club box, Annual Report 1905–06, 22–23.
24. *A History of Hyde Park Travel Club*, compiled by Ruth B. Lord, published in Honor of the Fiftieth Anniversary, Chicago, 1938. Chicago Historical Society Archives, Hyde Park Travel Club box.
25. *Annual Meeting Report*, March 23, 1914, 4. Chicago Historical Society Archives, Hyde Park Travel Club box.
26. *Arche Club*, History of the Arche Club documents, Chicago Historical Society Archives, Arche Club box.
27. "Sarah E.R. Fitzwilliam," passport application printed January 22, 2008, from www.Ancestry.com.
28. "Visited Many Odd Corners of Europe," *The Bloomington* (Illinois) *Pantagraph*, November 29, 1913.
29. Ryerson Library Accessions Book, December 15, 1917, "Gift of Mrs. Sarah E R Fitzwilliam," accession number 13561–14040.
30. *Bulletin of the Art Institute of Chicago,* 12, no. 1 (January 1918): 10–11; *Bulletin of the Art Institute of Chicago,* 12, no. 2 (February 1918): 24–25; *Bulletin of the Art Institute of Chicago,* 12, no. 6 (September 1918): 100; and discussion with Laurie Chipps, Catalog/Reference Librarian at the Ryerson Library July 2008.
31. Probate Court of Cook County records, Estate of S.E.R. Fitzwilliam, proof of last will and testament.
32. Probate Court of Cook County records, Claim to Dr. Romaine N. Douglass.
33. Probate Documents, Sarah Raymond Fitzwilliam, Court of Cook County, Richard J. Daley Center Archives, Chicago, Illinois.

Notes

Appendix Literature Review

1. Linda Lawrence Hunt, *Bold Spirit: Helga Estby's Forgotten Walk Across Victorian America* (New York: Anchor Books, 2003), xii.
2. Stephan Preskill and Robin Jacobvitz, *Stories of Teaching: a Foundation for Educational Renewal* (Upper Saddle River, NJ: Merrill Prentice Hall, 2001).
3. Greg Michie, *Holler If You Hear Me: The Education of a Teacher and His Students* (New York: Teachers College Press, 1999).
4. Elizabeth Edwards, *Women in Teacher Training Colleges, 1900–1960: A Culture of Femininity* (London: Routledge, 2001), 1.
5. Linda Lyman, *How Do They Know You Care? The Principal's Challenge* (New York: Teachers College Press, 2000).
6. Mary Cordier, *Schoolwomen of the Prairies and Plaines: Personal Narratives from Iowa, Kansas and Nebraska, 1860s-1920s* (Albuquerque: The University of New Mexico Press, 1992).
7. M. Homes and B. Weiss, *Lives of Women Public Schoolteachers: Scenes From American Educational History* (New York: Garland Publishing, 1995).
8. June Edwards, *Women in American Education, 1820–1955: The Female Force and Educational Reform* (Westport, CT: Greenwood Press, 2002).
9. Helen Horowitz, *The Power and Passion of M. Carey Thomas* (New York: Alfred A. Knopf, 1994).
10. Joyce Antler, *Lucy Sprague Mitchell: The Making of a Modern Woman* (New Haven, CT: Yale University Press, 1987).
11. A. McCluskey and E. Smith, *Mary McLeod Bethune: Building a Better World* (Bloomington, IN: Indiana University Press, 1999).
12. P. Foner and J. Pacheco, *Three Who Dared: Prudence Cradall, Margaret Douglass, Myrtilla Miner-Champions of Antebellum Black Education* (Westport, CT: Greenwood Press, 1984).
13. Jana Nidiffer, *Pioneering Deans of Women: More than Wise and Pious Matrons* (New York: Teachers College Press, 2000).
14. M. Crocco, P. Munro, and K. Weiler, *Pedagogies of Resistance: Women Educator Activists, 1880–1960* (New York: Teachers College Press, 1999).
15. M. Crocco and O. L. Davis (eds), *Building a Legacy: Women in Social Education 1784–1984* (Silver Spring, MD: National Council for the Social Studies, 2002); and *Bending the Future to their Will: Civic Women, Social Education, and Democracy* (Boston: Rowman and Littlefield, 1999).
16. Stephanie Van Hover, "Deborah Partidge Wolfe and Education for Democracy," *Theory and Research in Social Education* 31, no. 1 (Winter 2003): 105–131.
17. Jackie Blount, *Destined to Rule the Schools: Women and the Superintendency 1873–1995* (Albany: State University of New York Press, 1998).
18. Patricia Carter, *Everybody's Paid But the Teachers: The Teaching Profession and the Women's Movement* (New York: Teachers College Press, 2002), 3.
19. Crocco, Munro, and Weiler. *Pedagogies of Resistance*.

20. M. Belenky, B. Clinchy, N. Gold Berger, and J. Tarule, *Women's Ways of Knowing* (New York: Basic Books, 1986).
21. Thomas Dynneson, Richard Gross, and Michael Berson, *Designing Effective Instruction for Secondary Social Studies* (Upper Saddle River, NJ: Merrill Prentice Hall, 2003).
22. Theodore Sizer, *Horace's Compromise: The Dilemma of the American High School* (Boston: Houghton Mifflin, 1984).
23. C. Furay and M Salevouris, *The Methods and Skills of History* (Wheeling, IL: Harlan Davidson, Inc., 2000), 236.
24. Joan Scott (ed.), *Feminism and History* (Oxford: Oxford University Press, 1996).
25. Carol Gilligan, *In a Different Voice: Psychological Theory and Women's Development* (Cambridge, MA: Harvard University Press, 1982).
26. Nel Noddings, "The Gender Issue," *Educational Leadership* 49 (1991/1992): 68.
27. Peter Martorella, *Teaching Social Studies in Middle and Secondary Schools* (Upper Saddle River, NJ: Prentice Hall, 2001), 301.
28. Furay and Salevouris, *The Methods and Skills of History*, 238.
29. Mark Gilderhus, *History and Historians* (Englewood Cliffs, NJ: Prentice Hall, 1996).
30. Keith Windschuttle, *The Killing of History: How Literary Critics and Social Theorists are Murdering our Past* (New York: The Free Press, 1996), 2.
31. Joyce Appleby, Lynn Hunt, and Margaret Jacob, *Telling the Truth About History* (New York: W.W. Norton and Company, 1994).
32. H. Giroux and P. McLaren (eds), *Critical Pedagogy, the State, and Cultural Struggle* (New York: State University of New York Press, 1989).
33. Nancy Woloch, *Women and the American Experience* (New York: McGraw Hill, 1996), 168.
34. K. Carter and W. Doyle, "Personal Narrative and Life History in Learning to Teach," in *Handbook of Research on Teacher Education*, ed. J. Sikula (New York: Macmillan, 1996), pp. 120–142.
35. Preskill and Jacobvitz, *Stories of Teaching*, 3.
36. Blount, *Destined to Rule the Schools*.
37. Joel Spring, *The American School: 1642–1996* (New York: McGraw-Hill, 1997), 124.
38. C. Larson and K. Murtadha, "Leadership for Social Justice," in *The Educational Leadership Challenge: Redefining Leadership for the 21st Century,* ed. J. Murphy (Chicago: National Society for the Study of Education, 2002), 139.
39. Margaret Grogan, *Voices of Women Aspiring to the Superintendency* (Albany, NY: State University of New York Press, 1996).
40. Nel Noddings, *The Challenge to Care in Schools: An Alternative Approach to Education* (New York: Teachers College Press, 1992).
41. Robert Starrat, *Building an Ethical School: a Practical Response to the Moral Crisis in Schools* (Washington, DC: Falmer, 1997).

42. Grogan, *Voices of Women*, 177.
43. Linda Lyman, Diane Ashby, and Jenny Tripses, *Leaders Who Dare: Pushing the Boundaries* (Lanham, MD: Rowman & Littlefield Education, 2005).
44. Blount, *Destined to Rule the Schools*.
45. L. Smulyan, *Balancing Acts: Women Principals at Work* (Albany, NY: SUNY Press, 2000).
46. Carol Gilligan, *In a Different Voice: Psychological Theory and Women's Development* (Cambridge, Mass.: Harvard University Press, 1993 [1982]); Nel Noddings, *The Challenge to Care in Schools: An Alternative Approach to Education* (New York: Teachers College Press, 2005 [1992]); Mary Field Blenky et al., *Women's Ways of Knowing: The Development of Self, Voice and Mind* (New York: Basic Books, 1997 [1986]).
47. Blount, *Destined to Rule the Schools*, 169.
48. Madeleine Grumet, *Bitter Milk: Women and Teaching* (Amherst: The University of Massachusetts Press, 1988), 58.

BIBLIOGRAPHY

Acker, Sandra. "Feminist Theory and the Study of Gender and Education," *International Review of Education* 33 (1987): 419–435.

Acker, Sandra, ed. *Teachers, Gender and Careers.* New York: The Falmer Press, 1989.

Addams, Jane. *Twenty Years at Hull-House with Autobiographical Notes.* Boston: Bedford/St. Martin's, 1999.

Alliance Library System, "Sarah Raymond." http://www.alliancelibrarysystem.com/IllinoisWomen/files/mc/html1/raymon.cfm. http://www.alliancelibrarysystem.com/IllinoisWomen/files/mc/html1/raymond.htm

Allison, Clinton. *Present and Past: Essays for Teachers in the History of Education.* New York: Peter Lang, 1995.

Alpern, Sara, Joyce Antler, Elisabeth Israels Perry, and Ingrid Winther Scobie, eds. *The Challenge of Feminist Biography: Writing the Lives of Modern American Women.* Urbana: University of Illinois Press, 1992.

The Alumni Quarterly of the I.S.N.U., "Sarah Raymond," May 1914 and February 1918.

Antler, Joyce. *Lucy Sprague Mitchell: The Making of a Modern Woman.* New Haven, CT: Yale University Press, 1987.

Apple, Michael W, and James A Beane, eds. *Democratic Schools.* Alexandria, VA: ASCD, 1995.

Appleby, Joyce, Lynn Hunt, and Margaret Jacob. *Telling the Truth about History.* New York: W. W. Norton and Co., 1994.

Ascher, Carol, Louise DeSalvo, and Sara Ruddick, eds. *Between Women: Biographers, Novelists, Critics, Teachers and Artists Write About Their Work on Women.* Boston: Beacon Press, 1984.

Bailyn, Bernard. *On the Teaching and Writing of History.* Hannover, NH: University Press of New England, 1994.

Barry, Kathleen. "Toward a Theory of Women's Biography: from the Life of Susan B. Anthony." In *All Sides of the Subject: Women and Biography,* ed. T. Iles, 21–35. New York: Teachers College Press, 1992.

Beadle, Muriel. *The Fortnightly of Chicago: The City and its Women: 1873–1973.* Chicago: Henry Regnery Company, 1973.

Belenky, Mary Field, Blythe McVicker Clinchy, Nancy Rule Goldberger, and Jill Mattuck Tarule. *Women's Ways of Knowing*. New York: Basic Books, 1986.

Belenky, Mary Field, Lynne A. Bond, and Jacqueline S. Weinstock. *A Tradition That Has No Name*. New York: Basic Books, 1997.

Bellamy, Edward. *Looking Backward*. New York: Dover Publications, 1996.

Benham, Maenette, and Joanne Cooper. *Let My Spirit Soar! Narratives of Diverse Women in School Leadership*. Thousand Oaks, CA: Corwin Press, 1998.

Benjamin, Jules. *A Student's Guide to History*. Boston: Bedford Books, 1998.

Birkett, Dea. *Spinsters Abroad: Victorian Lady Explorers*. New York: Blackwell, 1989.

Blair, Karen. *The Clubwoman as Feminist: True Womanhood Redefined, 1868–1914*. New York: Holmes & Meier Publishers, 1980.

Bloom, Leslie. *Under the Sign of Hope: Feminist Methodology and Narrative Interpretation*. Albany: State University of New York Press, 1998.

Bloomington School Board Minutes Record Book, 1868–1899. Archives, Bloomington District 87, Bloomington, Illinois.

Blount, Jackie. *Destined to Rule the School: Women and the Superintendency 1873–1995*. Albany: State University of New York Press, 1998.

Board of Supervisors of Kendall County. *Atlas and History of Kendall County, Illinois*. Elmhurst, IL: Friendly Map & Publishing Company, 1941.

Bolman, Lee G., and Terrence E. Deal. *Reframing Organizations*. San Francisco: Jossey-Bass, 1997.

———. *Reframing the Path to School Leadership: A Guide for Teachers and Principals*. Thousand Oaks, CA: Corwin Press, 2002.

Brands, H. W. *The Reckless Decade: America in the 1890s*. New York: St. Martin's Press, 1995.

Brown, Victoria. *The Education of Jane Addams*. Philadelphia: University of Pennsylvania Press, 2004.

Bryson, Valerie. *Feminist Political Theory: An Introduction*. Basingstoke, Great Britain: Macmillan, 2003.

Buhle, Mari Jo. *Women and American Socialism, 1870–1920*. Urbana: University of Illinois Press, 1981.

Buhle, Mari Jo, and Paul Buhl. *The Concise History of Woman Suffrage: Selections from the Classic Work of Stanton, Anthony, Gage, and Harper*. Urbana: University of Illinois Press, 1978.

Butcher, Patricia Smith. *Education For Equality: Women's Rights Periodicals and Women's Higher Education 1849–1920*. New York: Greenwood Press, 1989.

Calhoun, Charles, ed. *The Gilded Age: Essays on the Origins of Modern America*. Wilmington: Scholarly Resources, 1996.

Callahan, Raymond E. *Education and the Cult of Efficiency: A Study of the Social Forces That Have Shaped the Administration of Public Schools*. Chicago: The University of Chicago Press, 1962.

Carr, Wilford, and Stephen Kemmis. *Becoming Critical: Education, Knowledge and Action Research*. London: Falmer Press, 1986.

Bibliography 173

Carter, Kathy, and Walter Doyle, "Personal Narrative and Life History in Learning to Teach." In *Handbook of Research on Teacher Education,* ed. J. Sikula, 120–142. New York: Macmillan, 1996.

Carter, Patricia. *Everybody's Paid But the Teacher: The Teaching Profession and the Women's Movement.* New York: Teachers College Press, 2002.

Clinton, Catherine. *The Other Civil War American Women in Nineteenth Century.* New York: Hill and Wang, 1984.

Commemorative Portrait and Biographical Record of Kane and Kendall Counties, ILL. Chicago: Beers, Leggett & Com., 1888. (Biography of Jonathan Raymond and John West Mason were written by Miss Sarah E. Raymond, Bloomington, Illinois.)

Conway, Jill. *The Female Experience in Eighteenth-and Nineteenth-Century America: A Guide to the History of American Women.* New York: Garland Publishing, Inc., 1982.

Cook, John. *An Appreciation of Richard Edwards.* Undated and no publisher listed, Milner Library, Illinois State University.

Cook, John, and James V. McHugh. *A History of the Illinois State Normal University.* Normal, IL: Pantagraph Printing, 1882.

Cordier, Mary. *Schoolwomen of the Prairies and Plains: Personal Narratives from Iowa, Kansas, and Nebraska, 1860s–1920s.* Albuquerque: The University of New Mexico Press, 1992.

Crocco, Margaret Smith, and O. L. Davis, eds. *Bending the Future to Their Will: Civic Women, Social Education, and Democracy.* Boston: Rowman and Littlefield, 1999.

———. *Building a Legacy: Women in Social Education 1784–1984.* Silver Spring, MD: National Council for the Social Studies, 2002.

Crocco, Margaret Smith, Petra Munro, and Kathleen Weiler. *Pedagogies of Resistance: Women Educator Activists, 1880–1960.* New York: Teachers College Press, 1999.

Cuban, Larry. *The Urban School Superintendency: A Century and a Half of Change.* Bloomington, IN: The Phi Delta Kappa Educational Foundation, 1976.

———. *How Teachers Taught: Constancy and Change in American Classrooms 1890–1980.* New York: Longman, 1984.

Davis, Allen. *American Heroine: The Life and Legend of Jane Addams.* New York: Oxford University Press, 1973.

DeBlasio, Donna. *Her Own Society: The Life and Times of Betsy Mix Cowles, 1810–1876.* PhD dissertation, Kent State University, 1980.

Degler, Carl. *At Odds: Women and the Family in America from the Revolution to the Present.* Oxford: Oxford University Press, 1980.

Delegard, Kirsten. "Women's Movements, 1880s–1920s." In *A Companion to American Women's History,* ed. Nancy A. Hewitt, 328–347. Malden, MA: Blackwell Publishing, 2005.

Desmond, Humphrey. *The A.P.A. Movement.* Washington: The New Century Press, 1912.

Deutsch, Sarah. *Women and the City: Gender, Space, and Power in Boston, 1870–1940*. New York: Oxford University Press, 2000.

Dewey, John. *Democracy and Education*. New York: Macmillan, 1916.

———. *The School and Society and the Child and the Curriculum*. Chicago: The University of Chicago Press, 1990.

Dickson, Elmer. *Original Purchasers of Public Domain Land in Kendall County, Illinois*. Chico, CA: Elmer Dickson, 1995.

———. *Kendall County Country & Village Schools*. Chico, CA: Elmer Dickson, 2001.

———. *Teachers of Kendall County*. Chico, CA: Elmer Dickson, 2001.

———. *Kendall County Pioneers 1828–1840*. Chico, CA: Elmer Dickson, 2004.

Dickson, Elmer, ed. *1860 Census of Kendall County: With Supplemental Genealogical Data*. Chico, CA: Elmer Dickson, 1993.

———. *1850 Census of Kendall County: Annotated With Supplemental Genealogical Data*. Chico, CA: Elmer Dickson, 2004.

Douglas, Davison. *Jim Crow Moves North: The Battle over Northern School Segregation, 1865–1954*. New York, NY: Cambridge University Press, 2005.

DuBois, Ellen. *Feminism and Suffrage: The Emergence of an Independent Women's Movement in America 1848–1869*. Ithaca: Cornell University Press, 1978.

Dunlap, Diane, and Patricia A. Schmuck, eds. *Women Leading in Education*. Albany: State University of New York, 1995.

Dynneson, Thomas, Richard Gross, and Michael Berson. *Designing Effective Instruction for Secondary Social Studies*. Upper Saddle River, NJ: Merrill Prentice Hall, 2003.

Edwards, Elizabeth. *Women in Teacher Training Colleges, 1900–1960: A Culture of Femininity*. London: Routledge, 2001.

Edwards, June. *Women in American Education, 1820–1955: The Female Force and Educational Reform*. Westport, CT: Greenwood Press, 2002.

Evan, Sara. *Born for Liberty: A History of Women in America*. New York: Free Press Paperbacks, 1997.

Falcone, Joan. *The Bonds of Sisterhood in Chicago Women Writers: The Voice of Elia Wilkinson Peattie*. PhD dissertation, Illinois State University, 1992.

Farren, Kathy, ed. *A Bicentennial History of Kendall County, Illinois*. Yorkville, IL: Kendall County Bicentennial Commission, 1976.

Fennema, Elizabeth, and M. Jane Ayer, eds. *Women and Education: Equity or Equality?* Berkeley, CA: McCutchan Publishing, 1984.

Flanagan, Maureen. *Seeing With Their Hearts: Chicago Women and the Vision of the Good City, 1871–1933*. Princeton: Princeton University Press, 2002.

———. *America Reformed: Progressives and Progressivisms 1890s–1920s*. New York: Oxford University Press, 2007.

Foner, Philip S., and Josephine Pacheco. *Three Who Dared: Prudence Crandall, Margaret Douglass, Myrtilla Miner-Champions of Antebellum Black Education*. Westport, CT: Greenwood Press, 1984.

Bibliography

Freed, John. *Educating Illinois: Illinois State University, 1857–2007.* Virginia Beach, VA: Donning Company Publishers, 2009.

Furay, Conal, and Michael J. Salevouris. *The Methods and Skills of History.* Wheeling, IL: Harlan Davidson, Inc., 2000.

Gardner, Howard. *Multiple Intelligences: The Theory in Practice.* New York: Basic Books, 1993.

———. *Leading Minds: An Anatomy of Leadership.* New York: Basic Books, 1995.

Gere, Anne. *Intimate Practices: Literacy and Cultural Work in U.S. Women's Clubs, 1880–1920.* Urbana: University of Illinois Press, 1997.

Gifford, Carolyn De Swarte, ed. *Writing Out My Heart: Selections from the Journal of Frances E. Willard, 1855–1896.* Urbana: University of Illinois Press, 1995.

Gifford, Carolyn De Swarte, and Amy Slagell, eds. *Let Something Good Be Said: Speeches and Writings of Frances E. Willard.* Urbana: University of Illinois Press, 2007.

Gilderhus, Mark. *History and Historians.* Englewood Cliffs, NJ: Prentice Hall, 1996.

Gilligan, Carol. *In a Different Voice: Psychological Theory and Women's Development.* Cambridge, MA: Harvard University Press, 1982.

Giroux, Henry A., and Peter McLaren, eds. *Critical Pedagogy, the State, and Cultural Struggle.* New York: State University of New York Press, 1989.

Glanz, Jeffrey. *Bureaucracy and Professionalism: The Evolution of Public School Supervision.* Rutherford: Fairleigh Dickinson University Press, 1991.

Glass, Thomas. *The History of Educational Administration Viewed Through its Textbooks.* Lanham, MD: Scarecrow Education, 2004.

Glesne, Corrine. *Becoming Qualitative Researchers: An Introduction.* New York: Longman, 1999.

Gordon, Ann, ed. *The Selected Papers of Elizabeth Cady Stanton and Susan B. Anthony.* New Brunswick, NJ: Rutgers University Press.

Gordon, Lynn. "Education and the Professions." In *A Companion to American Women's History*, ed. Nancy A. Hewitt, 227–249. Malden, MA: Blackwell Publishing, 2005.

Graves, Karen. *Girls' Schooling During the Progressive Era: From Female Scholar to Domesticated Citizen.* New York: Garland Publishing, 1998.

Grogan, Margaret. *Voices of Women Aspiring to the Superintendency.* Albany, NY: State University of New York Press, 1996.

———. "Laying the Groundwork for a Reconception of the Superintendency from Feminist Postmodern Perspectives." *Educational Administration Quarterly* 36 (2000): 117–142.

Grumet, Madeleine. *Bitter Milk: Women and Teaching.* Amherst: The University of Massachusetts Press, 1988.

Gutierrez, Rachel. "What is a Feminist Biography?" In *All Sides of the Subject: Women and Biography,* ed. Teresa Iles, 48–55. New York: Teachers College Press, 1992.

Harmon, Sandra. "The Voice, Pen and Influence of Our Women are Aboard in the Land: Women and the Illinois State University, 1857–1899." In *Nineteenth-Century Women Learn to Write*, ed. Catherine Hobbs, 84–102. Charlottesville: University Press of Virginia, 1995.

Harper, Charles. *Development of the Teachers College in the United States with Special Reference to the Illinois State Normal University*. Bloomington, IL: McKnight & McKnight Publishers, 1935.

Hasbrouck, Jacob. *History of McLean County Illinois*. Topeka: Historical Publishing Company, 1924.

Heilbrun, Carolyn. *Writing a Woman's Life*. London: Women's Press, 1989.

Hendricks, Wanda. *Gender, Race, and Politics in the Midwest: Black Club Women in Illinois*. Bloomington, IN: Indiana University Press, 1998.

Herbst, Jurgen. *And Sadly Teach: Teacher Education and Professionalization in American Culture*. Madison: University of Wisconsin Press, 1989.

Hewitt, Nancy A., ed. *A Companion to American Women's History*. Malden, MA: Blackwell Publishing, 2005.

Hicks, Edmund. *Hick's History of Kendall County*. Yorkville, IL: Record Print, 1927.

Historical Encyclopedia of Illinois and History of Kendall County. Chicago: Munsell Publishing Company, 1914.

Hoffman, Nancy. *Woman's True Profession: Voices from the History of Teaching*. Cambridge, MA: Harvard Education Press, 2003.

Holmes, Madelyn, and Beverly J. Weiss. *Lives of Women Public Schoolteachers: Scenes From American Educational History*. New York: Garland Publishing, Inc., 1995.

Horowitz, Helen. *The Power and Passion of M. Carey Thomas*. New York: Alfred A. Knopf, 1994.

Hundley, Rex. *History of Irving School*. Unpublished manuscript written and compiled for the centennial year 1957 by the school librarian, 1957. Archives, Bloomington District 87, Bloomington, Illinois.

Hunt, Linda. *Bold Spirit: Helga Estby's Forgotten Walk Across Victorian America*. New York: Anchor Books, 2003.

Huntley, Marilyn A., ed. *Newark Reminisces*. July 1976. Unpublished manuscript in the Newark, Illinois, public library.

Hurst, Homer. *Illinois State Normal University and the Public Normal School Movement*. Nashville: George Peabody College for Teachers, 1948.

Iles, Teresa, ed. *All Sides of the Subject: Women and Biography*. New York: Teachers College Press, 1992.

Illinois State Normal University. *The Index: Class Annual*. Normal, IL: Illinois State Normal University, 1892–1900.

Illinois State Normal University. *The Alumni Register, 1860–1927*. Normal, IL: Illinois State Normal University, 1927.

Jacobs, Jerry A., ed. *Gender Inequality at Work*. Thousand Oaks: Sage Publications, 1995.

Bibliography

Johnson, Tony. *Historical Documents in American Education*. Boston: Allyn and Bacon, 2002.

Journal of Proceedings of the Fifty-First Annual Meeting of the Illinois State Teachers' Association and Sections held at Springfield, Illinois on December 27–29, 1904. Springfield, IL: Illinois State Journal Co., 1905.

Journal of Proceedings of the Sixty-Sixth Annual Meeting of the Illinois State Teachers' Association held at Springfield, Illinois on December 29, 30, and 31, 1919. Springfield, IL: Illinois State Journal Co., 1920.

"Jubilee Anniversary Celebration of the Illinois State Normal University," *The Daily Pantagraph,* June 24, 1897.

Karier, Clarence. *The Individual, Society and Education: A History of American Educational Ideas*. Urbana: University of Illinois Press, 1986.

Katz, Michael. *Reconstructing American Education*. Cambridge, MA: Harvard University Press, 1987.

Kaufman, Polly. *Women Teachers on the Frontier*. New Haven: Yale University Press, 1984.

Kessler-Harris, Alice. "The Just Price, the Free Market and the Value of Women." *Feminist Studies* 142 (1988): 235–250.

Kinzer, Donald. An *Episode in Anti-Catholicism: the American Protective Association*. Seattle: University of Washington Press, 1964.

Knupfer, Anne. *Toward a Tenderer Humanity and a Nobler Womanhood: African American Women's Clubs in Turn-of-the-Century Chicago*. New York: New York University Press, 1996.

Kourany, Janet A., James P. Sterba, and Rosemarie Tong. *Feminist Philosophies: Problems, Theories and Applications*. Upper Saddle River, NJ: Prentice Hall, 1999.

Kozol, Jonathan. *Savage Inequalities*. New York: Crown Publishers, 1991.

Kridel, Craig, ed. *Writing Educational Biography: Explorations in Qualitative Research*. New York: Garland, 1998.

Larson, Colleen L., and Khaula Murtadha. "Leadership for Social Justice." In *The Educational Leadership Challenge: Redefining Leadership for the 21st Century,* ed. Joseph Murphy, 134–161. Chicago: National Society for the Study of Education, 2002.

Lather, Patricia. *Getting Smart: Feminist Research and Pedagogy With/in the Postmodern*. New York: Routledge, 1991.

Lawrence-Lightfoot, Sara, and Jessica Hoffmann Davis. *The Art and Science of Portraiture*. San Francisco: Jossey-Bass Publishers, 1997.

Lawyer, Marvin. *The Old Rural Schools of Kendall County*. Self-published undated article found in the Kendall County Historical Society.

Light, Ivan. *This Blooming Town: A Sketch of Bloomington, Illinois*. Bloomington, IL: Light House Press, 1956.

Loewen, James. *Lies My Teacher Told Me: Everything Your American History Textbook Got Wrong*. New York: Simon & Schuster, 1996.

Loomis, Burt. *The Educational Influence of Richard Edwards*. Nashville: George Peabody College for Teachers, 1932.

Lorence, James J. *Enduring Visions Readings.* Lexington, MA: D.C. Heath and Company, 1993.

Lyman, Linda. *How Do They Know You Care? The Principal's Challenge.* New York: Teachers College Press, 2000.

Lyman, Linda, Dianne Ashby, and Jenny Tripses. *Leaders Who Dare: Pushing the Boundaries.* Lanham, MD: Rowman & Littlefield Education, 2005.

Marshall, Helen. *Grandest of Enterprises.* Normal, IL: Illinois State Normal University, 1956.

———. "The Town and the Gown." *Journal of the Illinois State Historical Society* (Summer 1957): 141–167.

Martin, Theodora. *The Sound of Our Own Voices: Women's Study Clubs 1860–1910.* Boston: Beacon Press, 1987.

Martorella, Peter. *Teaching Social Studies in Middle and Secondary Schools.* Upper Saddle River, NJ: Prentice Hall, 2001.

Massmann, Priscila. *A Neglected Partnership: the General Federation of Women's Clubs and the Conservation Movement, 1890–1920.* PhD dissertation, The University of Connecticut, 1997.

Matile, Roger, ed. *150 Years Along the Fox: The History of Oswego Township Illinois.* Oswego, IL: Oswego Sesquicentennial Steering Committee, 1983.

Matthews, Jean. *The Rise of the New Woman: The Women's Movement in America, 1875–1930.* Chicago: The American Way, 2003.

McCarthy, Kathleen, ed. *Lady Bountiful Revisited: Women, Philanthropy, and Power.* New Brunswick, NJ: Rutgers University Press, 1990.

McCaul, Robert. *The Black Struggle for Public Schooling in Nineteenth-Century Illinois.* Carbondale: Southern Illinois University Press, 1987.

McClelland, Averil Evans. *The Education of Women in the United States: A Guide to Theory, Teaching and Research.* New York: Garland Publishing, 1992.

McCluskey, Audrey Thomas, and Elaine Smith. *Mary McLeod Bethune: Building a Better World.* Bloomington, IN: Indiana University Press, 1999.

McLean County Genealogical Society. *Illinois McLean County Cemeteries.* Normal, IL: McLean County Genealogical Society, 2001.

McLean County Historical Society. *Transactions of The McLean County Historical Society, Volume II.* Bloomington, IL: Pantagraph Printing and Stationery Co., 1903.

McMillan, James H., and Sally Schumacher. *Research in Education: A Conceptual Introduction.* New York: Longman, 1997.

McNergney, Robert F., and Joanne M. Herbert. *Foundations of Education: The Challenge of Professional Practice.* Boston: Allyn and Bacon, 1998.

McPheron, Linda. *A Historical Perspective of Career Patterns of Women in the Teaching Profession: 1900–1940.* PhD dissertation, Illinois State University, 1981.

Merriam, Sharan. *Qualitative Research and Case Study Applications in Education.* San Francisco: Jossey-Bass, 1998.

Michel, Sonya, and Robyn Muncy. *Engendering America: A Documentary History, 1865 to the Present.* Boston: McGraw-Hill, 1999.

Michie, Greg. *Holler If You Hear Me: The Education of a Teacher and His Students.* New York: Teachers College Press, 1999.

"Model School System in Bloomington," *The Sunday Chronicle,* December 29, 1901.

Morgan, Marilon. *A Voice from the Shadows: A Historical Educational Case Study of Julia Ann Christian.* PhD dissertation, Oklahoma State University, 2001.

Nash, Margaret A. *Women's Education in the United States 1780–1840.* New York: Palgrave Macmillan, 2005.

National Council for History Education. *Bradley Commission Report,* 1995.

National Council for the Social Studies. 2000.

Neumann, Anna, and Penelope L. Peterson, eds. *Learning from Our Lives: Women, Research, and Autobiography in Education.* New York: Teachers College, 1997.

Newark's Sesquicentennial History: A Community Proud of its Heritage 1835–1985. Yorkville, IL: Kendall County Record, Inc., 1985.

Nidiffer, Jana. *Pioneering Deans of Women: More than Wise and Pious Matrons.* New York: Teachers College Press, 2000.

Noddings, Nel. "The Gender Issue." *Educational Leadership* 49 (1991/1992): 65–70.

———. *The Challenge to Care in Schools: An Alternative Approach to Education.* New York: Teachers College Press, 1992.

O'Donnell, Ellen. *Women in the Superintendency: A Research Synthesis and Biographical Case Study.* PhD dissertation, State University of New York at Binghamton University, 2001.

Ogren, Christine. *The American State Normal School: An Instrument of Great Good.* New York: Palgrave, 2005.

Ornstein Allan C., and Daniel U. Levine. *Foundations of Education.* Boston: Houghton Mifflin, 2000.

Ourada-Sieb, Theresa. *Lucy Nevels: A Profile of Leadership.* PhD dissertation, University of Nebraska, 1997.

Pai, Young, and Susan A. Adler. *Cultural Foundations of Education.* Upper Saddle River, NJ: Merrill, Prentice Hall, 2001.

Perlmann, Joel, and Robert Margo. *Women's Work? American Schoolteacher, 1650–1920.* Chicago: The University of Chicago Press, 2001.

The Personal Narratives Group, ed. *Interpreting Women's Lives: Feminist Theory and Personal Narratives.* Bloomington, IN: Indiana University Press, 1989.

Philpott, Thomas. *The Slum and the Ghetto: Neighborhood Deterioration and Middle-Class Reform, Chicago, 1880–1930.* New York: Oxford University Press, 1978.

Postman, Neil. *The End of Education: Redefining the Value of School.* New York: Knopf, 1995.

Preskill, Stephan, and Robin Jacobvitz. *Stories of Teaching: a Foundation for Educational Renewal.* Upper Saddle River, NJ: Merrill Prentice Hall, 2001.

"Rags to Riches: City School System's First Century Marked by Growth, Spiced by Strife." *The Daily Pantagraph,* February 3, 1957.

Raymond, Sarah. *Fifth Annual Report of the Bloomington Public Schools for the Year Ending June 10th 1881.* Bloomington, IL: Bulletin Printing and Publishing, 1881.

———. *Sixth Annual Report of the Bloomington Public Schools for the Year Ending June 8, 1882.* Bloomington, IL: Bulletin Printing and Publishing, 1882.

———. *Seventh Annual Report of the Bloomington Public Schools for the Year Ending June 8, 1883.* Bloomington, IL: Bulletin Co., Printers and Publishers, 1883.

———. *Rules and Regulations, Manual of Instruction to Teachers, and Graded Course of Study of the Public Schools of Bloomington, Ill.* Bloomington, IL: Bulletin Printing Co., 1883.

———. *Eighth, Ninth, and Tenth Annual Reports of the Bloomington Public Schools for the Years 1884, 1885, and 1886.* Bloomington, IL: Leader Publishing Co., Printers, 1886.

———. *Eleventh Annual Report of the Bloomington Public Schools for the Year 1886 and '87.* Bloomington, IL: Leader Publishing Co., Printers, 1887.

———. *Twelfth Annual Report of the Bloomington Public Schools for the Year 1887 and 1888.* Bloomington, IL: Leader Publishing Co., 1888.

———. *Thirteenth Annual Report of the Bloomington Public Schools for the Year 1888–89.* Bloomington, IL: Pantagraph Printing and Stationery Co., 1890.

———. *Fourteenth Report of the Bloomington Public Schools for the Year 1889–90.* Bloomington, IL: Pantagraph Printing and Stationery Co., 1891.

———. *Fifteenth Annual Report of the Bloomington Public Schools for the Year 1890–1891.* Bloomington, IL: The Leader, Printers, 1891.

———. *Sixteenth Annual Report of the Bloomington Public Schools for the Year 1891–1892.* Bloomington, IL: The Leader, Printers, 1892.

Raymond Fitzwilliam, Sarah. "History of the Public Schools of Bloomington: From 1825–1892." In *Transactions of the McLean County Historical Society: School Record of McLean County and Other Papers, Vol II*, Bloomington, IL: Pantagraph Printing and Stationery Co., 1903.

———. "The Heroic in Student Life." In *Semi-centennial History of the Illinois State Normal University*, ed. David Felmley, 222–224. Normal, IL: Illinois State Normal University, 1907.

———. "An Historical Review of the Bloomington High School." *The Alumni Aegis* 11 (1907): 127–129.

Read, Phyllis J., and Bernard L. Witlieb. *The Book of Women's Firsts.* New York: Random House, 1992.

Reed, Christopher. *Black Chicago's First Century: Volume 1, 1833–1900.* Columbia: University of Missouri Press, 2005.

Regan, Helen, and Gwen Brooks. *Out of Women's Experience: Creating Relational Leadership.* Thousand Oaks, CA: Corwin Press, 1995.

Reller, Theodore. *The Development of the City Superintendency of Schools in the United States.* Philadelphia: Self-published, 1935.

Reukauf, Diane. *The Mother's Voice in the Progressive Era: The Reform Efforts of Kate Waller Barett.* PhD dissertation, University of Virginia, 1994.

Riley, Glenda. *Inventing the American Woman: an Inclusive History, Vol. 2 since 1877.* Wheeling, IL: Harlan Davidson, Inc., 1995.

Rinehart, Alice, ed. *One Woman Determined to Make a Difference.* Bethlehem: Lehigh University Press, 2001.

Robertson, Stacey M. *Parker Pillsbury: Radical Abolitionist, Male Feminist.* Ithaca: Cornell University Press, 2000.

Rockefeller, Steven. *John Dewey: Religious Faith and Democratic Humanism.* New York: Columbia University Press, 1991.

Ryan, Kevin, and James M. Cooper. *Those Who Can, Teach.* Boston: Houghton Mifflin, 2000.

Scarlette, Erma. "A Historical Study of Women in Public School Administration from 1900–1977." EdD dissertation, The University of North Carolina at Greensboro, 1979.

Schlenker, Charlie. *Voices From the Past: Discovering Evergreen Cemetery.* Bloomington, IL: McLean County Historical Society, 2000.

The Schoolmaster. Bloomington, Illinois (May 1868–January 1871).

Scott, Anne. *Making the Invisible Woman Visible.* Urbana: University of Illinois Press, 1984.

———. *Natural Allies: Women's Associations in American History.* Urbana: University of Illinois Press, 1991.

Scott, Joann Wallach, ed. *Feminism and History.* Oxford: Oxford University Press, 1996.

Scott, Mary. *Annie Webb Blanton: First Lady of Texas Education.* PhD dissertation, Texas A&M University, 1992.

Seigfried, Charlene. *Pragmatism and Feminism.* Chicago: The University of Chicago Press, 1996.

Seigfried, Charlene, ed. *Feminist Interpretations of John Dewey.* University Park, PA: The Pennsylvania State University Press, 2002.

Semi-Centennial History of the Illinois State Normal University. Normal, IL: 1907.

"Semi-Centennial of High School." *The Pantagraph,* June 27, 1907.

Shakeshaft, Charol. *Women in Educational Administration.* Newbury Park, CA: Sage Publications, 1987.

Shank, Gary. *Qualitative Research: A Personal Skills Approach.* Upper Saddle River, NJ: Merrill, Prentice Hall, 2002.

Sizer, Theodore. *Horace's Compromise: The Dilemma of the American High School.* Boston: Houghton Mifflin, 1984.

Sklar, Kathryn Kish. *Florence Kelley and the Nation's Work: The Rise of Women's Political Culture, 1830–1900.* New Haven: Yale University Press, 1995.

Smith, Joan. *Ella Flagg Young: Portrait of a Leader.* Ames: Educational Studies Press and the Iowa State University Research Foundation, 1979.

Smith-Rosenberg, Carroll. "Politics and Culture in Women's History." *Feminist Studies* 61 (1980): 55–64.

Smith-Rosenberg, Carroll. *Disorderly Conduct. Visions of Gender in Victorian America.* New York: Alfred A. Knopf, 1985.

Smulyan, L. *Balancing Acts: Women Principals at Work.* Albany, NY: SUNY Press, 2000.

"Some Notes of His School History: How the School Has Grown, Capt. Burnham recalls interesting facts about the earlier years of organization," *The Daily Pantagraph,* December 3, 1914.

Solomon, Barbara Miller. *In the Company of Educated Women: A History of Women and Higher Education in America.* New Haven: Yale University Press, 1985.

Sorensen, Mark. "Ahead of Their Time: A Brief History of Woman Suffrage in Illinois," from http://www.historyillinois.org/suff.html. http://www.historyillinois.org/Links/Illinois_History_Resource_Page/suff.html.

Spring, Joel. *American Education.* New York: Longman, 1989.

———. *The American School: 1642–1996.* New York: McGraw-Hill, 1997.

Stanley, Liz. "Biography as Microscope or Kaleidoscope? The Case of 'Power' in Hannah Cullwick's Relationship with Arthur Munby." *Women's Studies International Forum* 101 (1987): 19–32.

Stanton, Elizabeth Cady. ed. *History of Woman Suffrage.* New York: Arno Press, 1969.

Starratt, Robert J. *Building an Ethical School: A Practical Response to the Moral Crisis in Schools.* Washington, D.C.: Falmer, 1997.

Stearns, Peter, Peter Seixas, and Sam Wineburg, eds. *Knowing, Teaching, and Learning History: National and International Perspectives.* New York: New York University Press, 2000.

Strober, Myra H., and David Tyack. "Why Do Women Teach and Men Manage? A Report on Research on Schools." *Signs: Journal of Women in Culture and Society* 5 (1980): 494–503.

Struck, Carol. *A Study of Staff Perceptions of the Source for Instructional Leadership in a Central Illinois School District.* PhD dissertation, Illinois State University, 1987.

Stuart, M. "Making the Choices: Writing about Marguerite Carr-Harris." In *All Sides of the Subject: Women and Biography,* ed. T. Iles, 59–67. New York: Teachers College Press, 1992.

Sullivan, Dolores. *William Holmes McGuffey: Schoolmaster to the Nation.* Cranbury, NJ: Associated University Presses, 1994.

Tompkins, Dortha. *District Eighty-Seven, Bloomington, Illinois.* Unpublished manuscript, 1976. Archives, Bloomington District 87, Bloomington, Illinois

Tong, Rosemarie. *Feminist Thought.* Boulder, CO: Westview Press, 1998.

Tozer, Steven, Guy B. Senese, and Paul C. Violas. *School and Society: Historical and Contemporary Perspectives.* Boston: McGraw-Hill, 2006.

Trofimenkoff, Susan Mann. "Feminist Biography." *Atlantis* 10, no. 2 (1985): 1–7.

Tyack, David, and Elisabeth Hansot. *Managers of Virtue: Public School Leadership in America, 1820–1980.* New York: Basic Books, 1982.

———. *Learning Together: A History of Coeducation in American Schools.* New Haven: Yale University Press, 1990.

Tyack David, and Larry Cuban. *Tinkering Toward Utopia: A Century of Public School Reform.* Cambridge: Harvard University Press, 1995.

Uglow, Jennifer. *Dictionary of Women's Biography.* London: Macmillan, 1982.

Urban Wayne J., and Jennings L. Wagoner. *American Education: A History.* Boston: McGraw-Hill, 2000.

Van Hover, Stephanie D. "Deborah Partidge Wolfe and Education for Democracy." *Theory and Research in Social Education* 31, no. 1 (Winter 2003): 105–131.

Wade, Louise Carroll. *Chicago's Pride: the Stockyards, Packingtown, and Environs in the Nineteenth Century.* Urbana: University of Illinois Press, 1987.

Wallace, Les. *The Rhetoric of Anti-Catholicism: The American Protective Association, 1887–1911.* New York: Garland Publishing, 1990.

Walton, Andrea, ed. *Women and Philanthropy in Education.* Bloomington, IN: Indiana University Press, 2005.

Ware, Susan. *Modern American Women: A Documentary History.* New York: McGraw-Hill, 1997.

Warren, Donald, ed. *American Teachers: Histories of a Profession at Work.* New York: Macmillan, 1989.

Wayne, Tiffany. *Women's Roles in Nineteenth-Century America.* Westport, CT: Greenwood Press, 2007.

Webb, L. Dean, Arlene Metha, and K. Forbis Jordan. *Foundations of American Education.* Upper Saddle River, NJ: Merrill, 2000.

Weiler, Kathleen. *Women Teaching for Change: Gender, Class, and Power.* New York: Bergin and Garvey, 1988.

Weimann, Jeanne. *The Fair Women.* Chicago: Academy, 1981.

Wheeler, Adade. *The Roads They Made: Women in Illinois History.* Chicago: Charles Kerr Publishing Company, 1977.

Wilmore, Elaine. *Principal Leadership.* Thousand Oaks, CA: Corwin Press, 2002.

Wilson, Margaret. *The American Woman in Transition: The Urban Influence, 1870–1920.* Westport, CT: Greenwood Press, 1979.

Windschuttle, Keith. *The Killing of History: How Literary Critics and Social Theorists are Murdering our Past.* New York: The Free Press, 1996.

Woloch, Nancy. *Women and the American Experience.* New York: McGraw-Hill, 1996.

Wood, Sharon E. *The Freedom of the Streets: Work, Citizenship, and Sexuality in a Gilded Age City.* Chapel Hill: The University of North Carolina Press, 2005.

Woyshner, Christine. "Race, Gender, and the Early PTA: Civic Engagement and Public Education, 1897–1924." *Teachers College Record* 105 (April 2003): 520–544.

Young, Michelle, and Linda Skrla, eds. *Reconsidering Feminist Research in Educational Leadership.* Albany: State University of New York Press, 2003.

Zinsser, Judith. *History & Feminism.* New York: Twayne Publishers, 1993.

Newspapers (Daily and Weekly)

Bloomington (Illinois) *Bulletin*
Bloomington (Illinois) *Leader*
Bloomington (Illinois) *Pantagraph*
Chicago (Illinois) *Tribune Chicago* (Illinois) *Post*
Illinois State Normal University Vidette

INDEX

Abolitionists, 20, 24–7, 36, 37
Act of 1857, 36
Adams, Jane, 33, 125, 126, 129
African American, 68, 69–72, 105, 117
All Around Club, *see* Boston Clubs
American Academy of Political and Social Science of Philadelphia, 129
American Equal Rights Association, 15
American Patriotic Association, 121
APA, *see* American Patriotic Association
Art Institute of Chicago, 6–7, 77, 129, 131

Beecher, Catherine, 32
Blackburn University, 88
Blackwell, Henry, 17
Bloomington Benevolent Society, 126
Bloomington High School, 31, 56, 67, 72, 73, 79, 84, 86, 87, 89, 90, 91, 94, 95, 96, 97, 99, 101, 103, 105
Bloomington-Normal school integration, 69–72
Bloomington Number 4 School, *see* Number 4 School
Bloomington Number 5 School, *see* Number 5 School
Bloomington Public Library, 132, 133, 146
Bloomington School Board elections, 15, 75, 115, 117–21
Bloomington School District 87, *see* District 87
Bloomington Teachers Association, 73, 74, 87
bookkeeping inaccuracies in school administration, 89
Boston Clubs
   All Around Club, 128
   National Folklore Society, 129
   Woman's Educational Society, 129
Boston Marriage, 79
Bridgewater Normal School, 46, 58
Brooks, Mary, 45
Brown, Prof. E. N., 114, 118. 119
Burnham, J. H., 19, 20, 28, 58
Burrill, Dr. Thomas J., 58

Capen, Charles, 7, 165 (note)
Captain F. J. Fitzwilliam, *see* Fitzwilliam, F. J.
case study methodology, 11, 137, 138, 142, 145, 147, 150
Catholic vs. Protestant teachers controversy, 16, 116–17, 120–2
Centennial Exposition in Philadelphia, 92–3
Central Illinois State Teachers Association, 13
Central Illinois Teachers Association, 126
Chase v. Stephenson, 71–2

Chicago Club Memberships
    Arche Club, 131
    Chicago Women's Club, 131
    Travel Club, 130–1
Chicago Historical Society, 131
Civil War and women's opportunities, 14–15, 32
coal mine visit, 57
colored, see African American
common school, 41, 44, 108, 148
compulsory education, 74, 122
Congregational Church, 24–5, 29, 88
Cook, John, 49, 56, 96
constructivism, 138
corporal punishment, 89, 92, 105
curriculum contributions of Sarah Raymond, see Raymond, Sarah; curriculum
curriculum, Illinois State University (ISNU), see Illinois State University (ISNU) curriculum

Daughters of the American Revolution, 6, 130
DAR, see Daughters of the American Revolution
DeGarmo, Dr. Charles, 58
demographics, 1891 high-school, 102
Department of Superintendence of the NEA (National Education Association), 80
Dewey, John, 125–6, 129
District 87, 3, 68, 146
Dunn, Harriet E., 61–2, 73, 79, 86, 87, 92, 105, 106

educational leadership, women in, 2–3, 9, 11, 12, 63, 137, 139, 146, 148–51
Edwards, Richard (2nd President of Illinois State Normal University), 36, 37, 41–5, 46–8, 52, 53–8, 64, 73, 88, 107, 122, 127
Etter, Samuel M., 57, 69, 70, 86, 90
Evergreen Cemetery, 3–4, 133

Fell, Jesse, 37, 84
feminist biography, 1–3, 125, 137–43, 151
feminist history/theory, 2, 140–3
feminist leadership theory, 148–51
Fitzgerald, Nellie, 103
Fitzwilliam, Captain F. J., 4, 6, 125, 129
Fitzwilliam, Sarah Raymond, see Raymond, Sarah
Fogarty, Katie, 103
foreign language instruction, 41, 99, 102, 105
Fowler Institute (Newark, Illinois), 31, 67
Free School Law of 1855, 56
frontier educators, 29–30, 58

Gaylord, S. D., 76, 80, 84, 85, 86, 89
Gardiner, Mary, 99
Gastman, Enoch, 58
gender in education
    equality issues, 51, 65
    male/female representation, 40, 73, 79, 96, 102, 113–14, 127
    salary discrepancy, 52, 86, 89, 90, 92, 99, 101–2, 104–5
    leadership style, 148–51
graduation ceremonies and speeches, 53–4, 74, 79, 96, 100, 106

Hale, Susan E., 73, 79, 86
Hayden, Hattie, 103
Herbartianism, 126
Hewett, Professor Edwin, 40, 58, 87, 88, 90
historiography, 142–5
Hovey, Charles (first Principal of Illinois State Normal University), 36, 45, 47, 58
Howe, Julia Ward, 17, 128
Howland, George, 100
Hull House, 125, 126
Hull, John, 79, 91

INDEX

Illinois State Normal University (ISNU)
   admission requirements, 38
   Alumni Association, 9–10, 63, 64–5
   co-education, 37–8, 49–50
   criticism, 52
   curriculum, 40–5
   history, 36–45, 72
   model school, 36, 38, 39, 40, 43, 45–9, 51, 55, 63, 65; *see also* model school concept
Illinois State Teachers Association, 13, 56
Illinois State University, *see also* Illinois State Normal University (ISNU)
   history, 45
   Milner Library, 146
IIlinois Wesleyan University, 13, 115, 126, 133
Industrial School and Home of Bloomington, 126
Industrial University (University of Illinois in Urbana/Champaign as it was later called), 50, 93
ISNU, *see* Illinois State Normal University (ISNU)

Jacoby, Jacob, 31, 75, 86, 94, 96, 118
James, Edmund J., 58
Johnson, Edith, 48

Kendall County, Illinois, 19, 21, 24–30, 35, 39, 128
Know-Nothingism, 121
Knox College, 24–5

Ladies Benevolent Association, 87
library board, 3, 78, 125
Lincoln, Abraham, 25, 37, 50, 104
Lyon, Mary, 32

Mann, Horace, 32, 46, 58
Marsh, Dr. B. P., 72, 73, 75, 79, 86, 90
Martha Crow, by her next friend v. Board of Education of Bloomington, 71

McGuffey Reader, 84
McLean County Museum of History, 68, 146
Metcalf School, 63
Metcalf, Thomas, 49, 52, 63, 64, 90
Methodist Church, 115, 125
Milner Library, *see* Illinois State University; Milner Library
model school concept, 46; *see also* Illinois State Normal University (ISNU); model school
Mott, Lucretia, 14
Moulton Hall, 63
Moulton, Hon. S. W., 58

National Folklore Society, *see* Boston Clubs
Native American encounters, 23, 28
NEA, Department of Superintendence, *see* Department of Superintendence of the NEA
Negro, *see* African American
"New Woman", 15–16, 130, 137, 146
night school, 93
Number 4 School, 99
Number 5 School, 69–70, 72

Oberlin College, 24, 27
obituaries of Sarah Raymond, *see* Raymond, Sarah; obituaries
Oneida colonies, 24–25

Pillsbury, W. L, 47, 48
Pike, Miss (a classmate of Sarah Raymond), 50, 61, 62
pioneer life and influence, 21–2, 23, 30, 62
Pierce, Cyrus, 46
Plato Club, 10, 125–6, 129
Pottawatomie Indians, 28
Progressive Era, 13, 126, 130,
Protestant vs Catholic teachers controversy *see* Catholic vs. Protestant teacher controversy

Raymond, Catherine, 4, 20, 21, 23, 28, 31
Raymond, Jonathan, 4, 9, 20, 21, 23, 24, 25, 28, 30
Raymond, Sarah
  Bloomington principal, 4, 12, 19, 26, 31, 69–79, 105, 111
  Bloomington teaching, 67–9
  in Chicago, 4, 5, 10, 24, 120, 122, 125, 129, 130–1, 132, 133–4
  club activities, 8, 10, 13, 64, 125, 126, 128, 129–31
  curriculum, 2, 13, 97, 99, 101, 107, 108–10, 138, 139
  early schooling, 28–30
  early teaching, 30, 67
  Manual of Instruction, *see* Rules and Regulations, Manual of Instruction to Teachers and Graded Course of Study of the Public Schools of Bloomington
  marriage to Captain F. J. Fitzwilliam, 129
  obituaries, 5–7, 9–11, 31, 32
  open door-open heart policy, 98
  personal accounts of Illinois State Normal University, 42, 59–62
  personnel practice, 98–9
  resignation, 113–123
  superintendent of schools, 4, 5, 6, 10, 12, 19, 26, 48, 51, 53, 64, 68, 75, 79–107, 113, 118, 122, 133
Republican Party, 24, 36–7, 84, 122, 128
Rider, Olive, 48
Religion of teachers, *see* Catholic vs. Protestant teachers controversy
Rudd, Helen, 51–2
Rules and Regulations, Manual of Instruction to Teachers and Graded Course of Study of the Public Schools of Bloomington, 97, 101, 108–11, 147

salaries of administrators, 69, 73, 86, 89, 92, 95, 99, 100, 102, 105, 118, 119

salaries of teachers, 69, 89, 90, 95, 99, 104
Sarah Raymond School, 3, 100–1
Sattley, Olive, 58
scarlet fever, 93
school administration, early history of, 80–1
school finances, 84–5, 86, 89, 90, 95, 99, 100, 102, 103, 104, 105
school integration, *see* Bloomington-Normal school integration
school maintenance issues, 88–9, 98
School Mistresses' Club of Illinois, 13, 126
smallpox, 93
Smith, William Hawley, 59
Stanton, Elizabeth Cady, 14, 17, 127
Stevenson, Adlai I, 3, 67, 80, 127–8
Superintendence, Department of, *see* Department of Superintendence of the NEA
superintendent of schools, role of, 84–5

Taft, Lorado, 77–8, 129
teacher contract terms, 100
teacher institutes, 55–7
teacher salaries, *see* salaries of teachers
temperance, 14, 15, 16, 116, 121
Thayer, G., 45
Tillinghast, Nicholas, 46, 58
Tipton, Judge Thomas F., 71, 74
training of school administrators, 80
Trotter Fountain, 10, 77
Trotter, Georgiana, 4, 75, 76, 77, 78–9, 89, 91, 98, 100, 104, 125, 129, 131–2, 147

underground railroad, 20, 24, 25, 26, 27–8
universal education, 14, 51
University of Illinois, *see* Industrial University of Illinois

vaccination, compulsory law, 94
Van Petten, Edwin M., 86, 114, 115, 119

Ward 5 School, *see* Number 5 School
WCTU, *see* Woman's Christian Temperance Union
Wilbur, Professor C. D., 30
Willard, Emma, 32
Willard, Frances, 116
Withers Park, 3, 77
Withers Public Library, 78, 79
Woman's Christian Temperance Union (WCTU), 16, 116, 117
Woman's Educational Association of Illinois Wesleyan University, 13
Woman's Educational Society, *see* Boston clubs
Woman's National Loyal League, 15
Woman's State Teachers' Association, 13, 126, 127
women's biography, *see* feminist biography
women's history/theory, *see* feminist history/theory
women's leadership theory, *see* feminist leadership theory
women's rights, 13–15, 17, 65, 76, 115, 130, 154 (note)
women's suffrage, 13–17, 51, 75, 115, 116, 127, 142, 144
Wrightonia Society, 49–52
    debate topics, 50–1

Young, Ella Flagg, 122, 126